PROBLEMS AND SOLUTIONS IN ZOOLOGY

Problems and Solutions in Zoology

ARNOLD C. LONG

CYBER TECH PUBLICATIONS

4264/3, Ansari Road, Daryaganj, New Delhi-110002 (India)

Ph.: 011-23244078. Fax: 011-23280028

E-Mail: cyberpublicationsdelhi@yahoo.com

Problems and Solutions in Zoology

First Edition 2009

Published by G.S. Rawat
For Cyber Tech Publications
4264/3, Ansari Road, Daryaganj, New Delhi-110002
Ph.: 011-23244078. Fax: 011-23280028
E-Mail: cyberpublicationsdelhi@yahoo.com

Printed at
Asian Offset Printers

Preface

Zoology is the study of life in its entirety. The growth of Zoology as a natural science during the last 1000 years is interesting from many points of view. One feature of this growth is changing empasis. Initially, it was description of life forms. Identification, nomenclature, classification of all recorded living forms, enjoyed the attention of scientists for a long time. Description of habitats of animals and their behaviour was included in this study.

Zoology is concerned with the study of animals and their living conditions. This book includes the Problems and Solutions of Zoology in a more detailed and descriptive way. This book would make the readers experience the humbling effect of this realisation that all organisms are related to each other virtue of sharing genetic material. We hope that this book would help the students to become more familiar with the diversity of animals.

We are indebted to the various authorities on the subject. We gratefully acknowledge them.

Last but not the least my publishers who appreciated this book and brought this work out in a record time.

—Editor

A

Aardwolf. Small, nocturnal, hyaena-like, carnlvorous mammal found in South Africa; feeds on termites.

Abactinal. Aboral: body surface opposite to oral or actinal side in starfishes, jelly-fishes.

Abdomen. Body region posterior to thorax. and containing viscera In mammals limited between diaphragm and pelvis and in invertebrates the region containing most of the digestive tract.

Abductor-Muscle. Muscle that draws a part away from axis of body or a limb or separates two parts e.g., in pedicillaraiae of star-fish where these muscles move apart the lateral calcareous pieces.

Abiogenesis. Theory given before the end of 17th century proposing that animal species arise spontaneously from non-living matter e.g., insects from dew, frogs and toads from muddy bottoms of ponds under the influence of sun, butterfies from cheese and fly-maggots from flesh.

Environment. Physical and chemical factors that influence living organisms.

Abomasum. Fourth chamber of stomach found in ruminants e.g. cow, sheep, goat etc., bearing longitudinal folds and gastric glands. Considered to be real stomach in these animals. the other regions of stomach are considered to be divisions of oesophagus.

Aboral. Body side opposite to mouth. Animals with bilateral symmetry have anterior end called oral end. In Jelly-fishes surface 'of body is directed upwards; in star-fishes abactinal surface.

Abortion. Operation to terminate pregnancy: premature delivery. giving out dead foetus.

Absorption. Taking of fiuids or other substances by tissues e.g. the products of digestion, vitamins. minerals and fluids across mucosa of stomach and intestine to enter the blood or lymph.

Abyss Mal. Ocean zone at depth of more than 2000 metres all aquatic zones below bathyal zone, also called deep sea; accounts for 87% of the ocean pressure increased to several hundred atmospheres, no light and total darkness. Fishes, protistans. worms. molluscs. sea-cucumbers and sea-stars are found in this zone.

Acanthopore. Tubular spine in some extinct bryzoans.

Acanthosome. Last primitive larval stage of certain prawns (Sergestidce: Crustacea).

Acanthozooid. Speclalised member of bryozoan colony which secretes tubules that appear as spines above the colony's outer surface.

Acarologist. Person having sound knowled~e of acarology.

Acarology. Branch of zoology that deals with study of mites. **ACCESSORY : chromosome.** sex-chromosome.

Acclimatisation. Adjustment of an animal/organism with some particular condition of its environment by bringing about necessary physical. physiological. biochemical and behavioural changes e.g., frog being cold-blooded animal cannot withstand excessive hot or cold weather; during winters it undergoes hibernation which involves lowering of BMR, consequent suspended animation thus minimising the vital needs of the body to cope with conditions of cold whereas during; summers it undergoes aestivation.

Accretion. Process of growth in which material is added to the outside in non-living matter .

Acellular. Having no cells or compartments e.g., Protozoa. The term has been coined considering the protozoan cell as a complete organism. Since a single cell of these animals, as compared to metazoan" cells, is capable of discharging all the functions of life and therefore considered equivalent to organism and not the cell.

Acetabulum. Cup-shaped socket of hip into which femur of hind limb fits.

Acetylcholine. (CH_3-CO-O-CH_2-CH_2-N$(CH_3)_3$-0H) acetic acid ester of. organic base choline; released from terminals of cholinergic axons of many neurons; responsible for transmission of a nerve impulse across a synapse; is hydrolysed to choline anc;I acetate by acetyl choline esterase which is present on po;st-synaptic membrane. The enzymatic destruction of substance kerminates its effects on post-synaptic membrane. The choline 'is reabsorbed actively by post-synaptic endings in which it is recycled by combination with acetyl CO-A to form a new molecule of the substance.

Acetylocholinesterase. Enzyme that hydrolyses acetyl-choline into choline and acetate; present on the surface of post-synaptic membrane. Enzyme molecule has two distinct sites - an anionic site that binds the quarternary nitrogen of acetylcholine and an esteratic site that can donate electron to the acetate portion of acelylacholine molecule cleaving it from the choline moiety. This allows the hydrolysis of acetylcholine and terminates the post-synaptic action of the transmitter.

Achromatic Figure. Metaphase plate chromosomal line up. *Achromic* : colourless: without normal pigmentation

Acid. compound which yields hydrogen ions to base in a chemical reaction; having pH below 7 *e.g.* HCI, H2S04; pH of a solution is merely the negative logarithm of hydrogen ion concentration herice a very low pH *e.g* . pH-l represents an extremely acidic solution; forms salts by reacting with bases.

Acidophil. Organism having preference for ardic environment. ACIDOSIS: pathogenic condition resulting from accumulation of acid or loss of base from the blood; characterised by increase of hydrogen ion concentratioll of blood or fall below 7.39 -7.45 in arterial and 7.39 - 7.42 in venous blood. May be caused by respiratory or metabolic cause. Defective pulmonary excretion of carbon dioxide results in increased blood H2C03 level causing a fall in blood pH. Decrease in blood bicarbonate level without change in H2C03 fraction due to various metabolic activities of the body also results in acidosis. Metabolic acidosis occurs very commonly in uncontrolled diabetes mellitus.

Acinus. One of the small terminal sacs in lung or multicellular gland.

Acoelous. First two centrusm of a vertebra where both the sides are convex *e.g.* IX vertebra of frog.

Acorn-Worm. Common name for hemichordate *Balanoglossus.*

Acguired Characters. Modifications of body that occur during life time of an individual. as a result of environmental conditions. For example. skin colour of human beings changes in different temperatures. These characters are heritable if involve some genetic change *viz* , mutation. otherwise are not inherited if involve only somatic change (as wrestler's body built).

Acraein. Protective and distasteful secretion. ejected by certain butterflies (depidoptera. Ht.xapoda)

Acrania. (*Acraniata*) : major groups of primitive chordates with no distinct cranium or skull like *Balanoglossus. Herdmania* and *Amphioxus.*

Acrocentric. Chromosome in which one arm is shorter than the other one. since centromere is close to one end.

Acromegaly. Disease caused by hypersecretion of anterior pituitary after bones have reached full growth; bones become thickened especially in the limbs and face; characterised by an overgrowth of mandibles. phalangeal cartilages and related areas, forehead. nose, hands. feet etc. Body becomes disproportionate giant.

Acromion. Process which is prolongation of spine of scapula forming the point of shoulder of mammls; present on outer side of coracoscapula and ending ventrally in an expanded acromial process. In four-footed mammals where clavicles are weakly developed or disappeared, a backward meta-acromian process is developed from acromial process.

Acron. Prostomial region of trochophore larva in some molluscs: unsegmented head of fossil Arthropods.

Acrosomal Cap. Covering over the head of a sperm derived from the golgo body.

Acrosome. Pointed tip of a sperm head that makes contact with an ovum at the time of fertillsation. It contains male pronucleus.

Acth. Adrenocorticetropic hormone: peptide secreted from anterior· pituitary lobe (adenohypophysis); acts on adrenal cortex to increase the synthesis and secretion of steroids in adrenal cortex leading to conservation reaction during stress. (See homeostasis). The corti co tropic releasing hormone(CRH) from hypothalamus stimulates release of the hormone whereas the hormone itself is responsible for inhibiting its release.

Actin. Muscle protein associated with I-bands of striped muscle fibres; filaments resemble two strings of beads twisted into a double helix. Each bead is molecule of G-actin (55 A) polymerized to form double helix of F-actin (fibre-like appearance). F-actin double helix has a pitch of about 710

A' so that its two strands cross over each other once every 355 A) Actin filaments are about Mm long and 80 A thick. They are joined at one end to material that constitues Z-ltne. Positioned in the grooves of actin helix are filamentous molecules of another protptn tropomyosin (see tropomyosin)

Actinal. Lower surface of the body in star-fishes; also called oral surface; characterised by presence of actinostame surrounding the mouth (See oral).

Actinaria. Order or subclass of Zoantharia (Actinoza: Coelenterata) that includes sea-anemones.

Actuiopodea. Subdass of Sarcodina (Protozoa: Protista); having highly specialised pseudopodia that help in free-floating.

Action Potential. Transient all-or-none reversal of a membrane potential produced by regenerating in current in excitable membranes; produced by membranes of neurones. muscles cells. some receptor cells, secretary cells and protozoa. The function being (1) rapid transmission of information over long distances in nerves and muscle fires and (2) to control the effect or esponses *e.g.* muscle contraction and neurotransmitter and holmonal release. Ranges in duration from less than a millisecond in axons to nearly a second in heart muscles cells.

When interval between two action potentials is reduced. the second one becomes progressively smaller and suddenly fails completely if initiated soon after end of first action potential which depends upon strength of stimulus.

Activation Energy. The kinetic energy required to cause two colliding molecules to react: energy required to bring reactant molecules to velocities sufficiently high to break or make chemical bonds; measured as the number of calories required to bring all the molecules in a mole of reactant at a given temperature to an activated state.

Activation Heat. Heat produced during excitation and activation of muscle tissue but independent of shortening.

Activator. Substance that can change inactive enzyme into active catalyst and bring about activation of chemical reaction. All proteolytic enzymes are secreted in inactive state and require some activator in order to prevent autolysis *ego* Pepsinogen is activated by acidity of stomach. Trypsiongen is activated by enterokinase of succus entericus.

Active Site. Catalytic region of enzyme molecule; special portion of enzyme surface to which substrate molecule 'fits'. The enzyme molecule has one or more peptide chains folded about forming tertiary structure of more-or-less globular protein of specific conformation. The active site is thought to consist of side groups of certain amino acid residues. which in the tertiary structure are brought into conformation to which substrate attaché & through electrostatic bonding. hydrogen bondinp;. Van der Wall's forces etc.

Active Transport. Energy requiring transportation of a substance across a membrane. usually against concentration gradient or electrochemical gradient. It requires ATP or other chemical substance containing bond energy. It can be selectively inhibited" by certain metabolic pathway blocking agents.

Acttomysoin. Complex of muscle protein and myosin; formed during muscle contraction; has higher viscosity than dissociated mixture of action and myosin; sensitive to changes in concentration of calcium. magnesium. potassium and hydrogen ion concentration and also ATP level; small amounts of ATP induce contraction of actomyosin.

Accupuncture. Old system of medical treatment involving piercing the body at specific points with specific number of metalic needles (particularly A,g. Au.,7n) t'\ particular pattern. to treat specific diseases.

Acyl Carrier Protein (ACP). Protein to which 'acyl' residues are attached during fatty acid synthesis.

Adaptive Convergence. Presence within a series of comparable ecological niches of only distantly related forms that superficially resemble qne another in morphological and other characters. correlated· with -very similar or identical environmental conditions.

Adaptive Radiations. Evolution and spread of a single line of descent of organisms into several ecological niches resulting in a series of strikingly different forms each adapted to a particular habitat *e.g.,* various mammals like. bat (flying). kangaroo (jumping) are evolved from a common five toed primitive insectivore mammal and during evolution the nature of limbs has been specialised to adapt to a variety of environments.

Adder. Small poisonous snake. *Vipera* spp. (Ophidia. Reptitia) about 60 cm long with black Zizag mark along the back. feeds mainly on lizards and mice.

Addax. Rare. large. light coloured (Khttish) antelope. *Addax nasomaculatus* (Artiodactyla: Mammalia) with spiral horns found in North Mrican deserts.' feeds on plants and can live almost indefinitely without drinking.

Addison's Disease. Atrophy of adernal glands by pathological. disorders or tubercular infection results in chronic insufficiency of cortisol and aldosterone secretion causing this disease; characterised by muscular weakness and ready faUgab1l1ty. pigmentation of skin and buccal mucosa loss of weight and dehydration; hypotension, anorexia, nal\sea. vomitting and diarrhoea'; mental confusion; hypoglycemia; decreased body hair growth and decreased ability to withstand infections. haemorrhage and other stresses (see adrenals. cordsols and aldosterone) .

Adductor Muscle. Muscle that draws part of body towards, median axis'; draws parts together *e.g.,* in mollusc *Lamellidens* species two shell-valves are drawn together by adductor muscles.

Adenine. Constituent of DNA; organic base of purine group. forms nucleotide by combining with deoxyribose sugar and phosphoric acid; associated with complementary base thymine forming cross-bar of hydrogen bond in the DNA helix (equation).

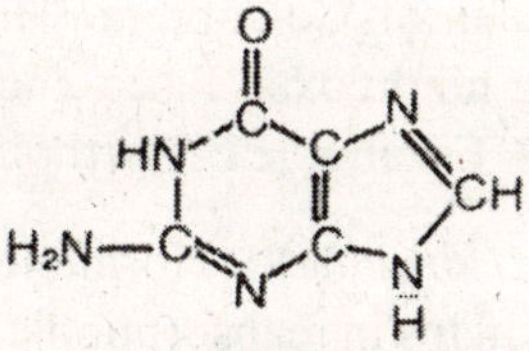

Adenine

Adenopophysis. Anterior pituitary gland or anterior lobe of pituitary; made up of interlacing cell cords and extensive network of sinusoids. Endothelium of sinusoids is fenestrated. The cells. contain granules of stored hormones which are extruded from cells by exocytosis: the granules have not been seen entering the capillaries and presumably dissolve in perisinusoidal space. It secretes six important hormones namely. thyroid stimulating hormone (TSHjthyrotropin), adrenocorticotropic hormone (ACTHjcorticotropin), growth hormone (GHjSTHjsomatotropin). follicle stimulating hormone (FSH), luteinising hormone (LH), interstitial cell stimulating hormone (ICSH}, prolactin (LTHjlactogenic hormonej luteotropinj mammotropl':).

Adenosine. Phosphorylated organic compound that functions in intracellular energy transfers.

Adenosine Triphosphate (ATP). Energy rich nucleotide used in cellular metabolism and activities required input of energy (endergonic reactions). When the substance is converted to ADP or AMP one or two phosphate groups are released giving up one or two energy-rich bonds to the reaction. Hydrolysis of first bond releases 3000 calories and that of second bond releases 12.000 cals.

Adenylate Cyclase. Enzyme associated with hormout: action at its target cell catalysing the hydrolysis of, ATP (adenosine triphosphate) to c-AMP when it is allostertccilly activated due to conformational changes in recc:ptor molecule present in target cell membrane which is brought about by interaction of receptor molecule. The compound is closely associated with receptor molecules but faces towards cytoplasm.

Adephaga. Suborder of Coleoptera (Hexapoda) that includes beetles having hind coxae fused and immovable: embraces super families Coraboidea. Gyrtnoidea and Paussoidea.

ADH (*Antidiuretic Hormone*). Hormone secreted by posterior pituitary lobe; helps in reabsorption of water.from glomerular filtrate and thus controls urine-formation (diuresis). In kidney tubule Na+ concentration around the proximal and distal convoluted tubules are in concentration state equivalent to that of most of body (300-350 milliosmols) but the concentration of Na+ ions at the turn of tubule'e loop of Henle and proximal to that is much higher (about 1500 milliosomols). This ionic concentrfation gradient pulls water out of collecting tubule and helps capillary water reabsorption. This whole mechanism is controlled by hormone. In hyposecretion of the hormone the Na+ ions are absorbed and thus concentration gradient is disturbed resulting in passing out large amounts of water in the urine causing a pathological condition known as Diabetesinsipidus.

Adipose. Fat or tissue in which fat is stored: type of connective tissue in which cells are rounded or polygsmal with thin layers of cytoplasm and nucleus on one side. These contain droplets of fats which may form larger granules. Pathogenic condition, adiposis. refers to deposition fat in the body see obesity. *Tissue.* Cells in animal body. meant for storage of lipid.

Adipocere. Fatty substance resulting from decomposition of animal bodies in moist places or under water.

Adolescence. Period of life-cycle of organism during which quiscent reproductive apparatus develops and becomes full) functional. In man it begins at the age of twelve years and in. women at eleven. *Drenal Cortex.* Outer part of adrenal gland; divided into three zones of variable distinctness - outer zona glomerulosa made up of whorls of cells which are continuous with columns of cells of middle zone called zona fasciculata (these columns are separated by venous sinuses). innermost zona reticulari where all the columns become interlaced in network. The cells contain abundant lipids which are specially more in outer portion of zona fasciculata. It secretes glucocorticoids with a widespread effects on metabolism of carbohydrates and proteins. mineralocorticoids with essential role in maintenance of sodium balance and extra cellular fluid volume and sex hormones which exert minor effects on the reproductive functions. *Glands.* Two small endocrine glands perched like cocked hats on the top of kidneys; each has two parts - outer cortex and inner medulla. Cortex secretes glucocorticoids. mineralocorticoids and sex hormones while medulla secretes two catecholomanes namely. epinepherine and nor-phertne.

Adrenal Medulla. Inner portion of adrenal gland; made up of interlacing cords of densely intervated granule-containing cells. Two types of cells can be' distinguished morphologically -epinephrine secreting cells, (having dense granules) and nor-epinephrine secreting cells (in which the smaller dense granules fail to fill the vesicles in which they are contained). Paraganglia-small groups resembling those of medulla are found near the thoracic and abdominal sympathetic ganglia. Medulla secretes epinephrine (adrenaline) and n6r-epinephrine (nor-adrenaline) which are collectively called catecholam1nes.

Adrenaline. Epinephrine: catecholamine synthesized from adrenal medulla; binds to p-adrenegric receptors. The hormone contributes to sympathetic response during emergencies. It increases heart rate. cardiac output, blood pressure. metabolism. oxygen consumption and blood sugar; stimulates respiration. brings about dilation of muscle vessels bronchus; vasocontriction of kidney and skin vessels' decrease total peripheral resistance.

Adrenergic Receptor. Membrane receptor molecules specific to interact with catecholamines; type of nerve fibre that releases adrenaline like substance from axon terminal during, nerve impulse transmission across the synapses. Two types of adrenergic receptors (a-adrenergic receptors and adrenergic receptors) can be distinguished by use of drugs that activate one or the other receptor types, eliciting responses characteristic of one or the other type of catecholamine.

α-Adenergic Receptors. The adrenergic receptors to which nor-adrenaline (nor-epinephrine) binds most effectively. ***β-Adrenergic Receptors.*** The adrenergic· receptors to which adrenaline (epinephrine) binds most effectively.

Adsorption. Adhesion of a very thin layer of liquid, dissolved substance or gas to a solid surface.

Adventitia. An outer layer surrounding some structure; dense fibrous connective tissue layer surrounding alimentary canal in neck and thorax region where serosa is not present.

Aedeagus. Copulatory organ or penis of male insects. ***Aepyornithidae.*** single family of extinct avian order Aepyornithiformes (Palaesgnathae:- Neornithes) - elephant birds etc.

Aepyornithiformes. order of extinct monofam1l1ar ratite birds (Aves: Chordata).

Aequorin. Animal protein extracted from the jelly-fish *Aequorea;* upon combining with calcium ions it emits a photon of bluish- green colour. The protein is used to measure the Ca++ in various tissues and body regions photocolorimetrtcally.

Aerial. Antmal/ organism inhabiting or frequenting air - birds and insects.

Aeriobic. Organism able to live and grow only in presence of free oxygen; processes that can orJy occur when oxygen is available *e.g.*, conversion of pyruvic acid to acetic acid before Krob's cycle.

Aerobiology. Scientific study of airborne micro-organisms. *Aerobiosis*. Life in presence of oxygen.

Aerophagia. Nervous persons. who hyperventilate, sometimes swallow large amount of air and some air swallowed is regurgitated (belching) and some of gases, it contains are absorbed but most of the air passes to colon and is expelled as flatus.

Affinity. Relationship of a group of animals/plants to the other group based upon similarities and differences of taxonomic characters to give a correct idea of its position in the animal/ plant kingdom.

Agama. Group of thick-tongued lizards (Agamidre Saurla) of Old World, about 30 cm long, varying in colour from brown, bl'le to orange and having aerodont dentition.

Agameon. Organism which reproduces only by asexual means. *Agamete*: etc asexual reproductive cell which develops into adult. *Agami*: golden-breasted tumpeter. *Psophr.a.* species (Aves) of South America which makes loud trumpet like sound.

Agglutination. Clumping or cohesion of red blood copuscles resulting from agglutinin-agglutinogen reaction (during mismatched blood transfusion when an 2.gglutinin reacts with agglutionogen and results in death of the patient).

Agglutinogen. A protein present in red blood corpuscles of human blood which causes production of specific agglutinls; a foreign substance, usually protein that causes production of specific antibody within an animal body. In human being two types of agglutinogens are recognised - A and B which may induce production of b and a agglutinIns respectively if blood transfusion is mismatched (see blood groups).

Aggressive Mimicry. Mimicry adapted by certain animals to attract or deceive others in order to prey upon them.

Aglaspide. Fossil order of arthropod animals (Xiphosurinda) characterised by phosphatic exoskeleton and trHobbed body.

Agnatha. Class of phylum Chordata that includes fish-like vertebrates having long slender body; skin smooth and sliming without any scales. no Jaws but sucker-like buccal funnel at anteriot end - hagflshes. lampreys etc.

Alar-membrane. Prepatagium: fold of skin between upper and fore-ann of bird's wing.

Albatross. The largest oceanic-bird (sl. *Diomedea* spp. (Procellaricormes: Aves) allied to petrel, having wing-spread of more than four metres;)

Albinism. Absence of pigmentation from skin. hair and eyes, inborn defect The tyrosinase enzyme catalyses transformation of tyrosine to dopamine which is converted to melanin; defect is caused by genetic deficiency due to non-formation of enzyme tyrosinase hence dopamine is not synthesised., so the skin receives no pigment. Recessive in nature and inherited like typical Mendelian recessive character.

Albino. Individual lacking body pigmentation - white rats and albino human beings.

Albumen. Protein synthesised in oviduct (specially in birds) from . amino acid removed from blood, called white of egg divisible lnto dense and thick albumen, later forming chalazae in bird egg when it rolls down through oviduct. Concerned with nutrition, osmoregulation of bird egg during development.

Albumins. Simple proteins (yielding only amino acids on hydrolysis) soluable in water and can be precipitated by saturated salt solutions; denatured on heating too much; coagulate on heating at low temperature (the property being useful in their detection). In natural state these are frequently associated with small quantities of mucoplysaccharides - serum albumin (blood serum). lactalbumin (milk), egg albumin (white of egg) etc.

Alcatras. Common name for large water-birds, such as gannets, peticans and others.

Alchoholic Fermentation. Form of anaerobic respiration in which glucose is broken down to form ethyl alcohol and CO_2 ; carried out by yeast and vari6us other fungi and by certain bacteria; occurs outside the organism and catalysed by zymase secreted either by living cells or released on their death; usually stops due to cell poisoning when alchohol level reaches about 15%; process is used in brewing, wine making, baking etc.

Alkalosis. Patholqglcakcondition of body resulting from accumulated ion of base or loss of acid from blood; characterised by increased pH above 7.39-7.45 in arterial and 7.39-7.42 in venous blood. May be caused due to respiratory or metabolic disturbance; excessive loss of CO_2 from blood lowers H_2C0_3 level and increases blood pH leading to respiratory alkalosis. Accumulation of bicarbonates in blood results in metabolic alkalosis.

Alkaptoneuria. Metabolic disorder involving block in chain of reactions by which amino acids are converted into simpler compounds; as a result an intermediate compound collects and darkens urine and various soft parts of body.

Allantochorion. Specialised vascular extraembryonic tissue formed in eggs of birds and mammals by fusion of extra-embryonic visceral mesoderm of allantois and chorion; helps in respiratory and excretory functioning of allantois.

Allantoin. In most animals (dogs etc.). uric acid formed during protein degradation is mddised to fonn it and then it is excreted in urine.

Allantois. Extraembryonic membrane arising as an outgrowth of cloaca in embryo of reptiles. birds and mammals'; consists of extraembryonic visceral mesoderm and extra-embryonic endoderm from tail end of embryo. former fusing with mesoderm of chorion to form allantochorion. It is supplied by allantoic artery carryi.lg deoxygenated blood to allantochorion where it Is oxygenated and returned via allantoic vein. CarrJes out respiratory and excretory functions.

Allergy. Strictly means an altered capacity to react; introduced by Von Pirquet in 1906 as specifically induced reactions which werc either enhanced or depressed; nowadays mostly used to mean altered reactivity which is increased above normal.

Alligator. One of the largest livign reptile (Crocodilia) Reptila) about three metres long havign short and broad snout with unequal teeth; carinvorous; formed in China and warm southern region of North America.

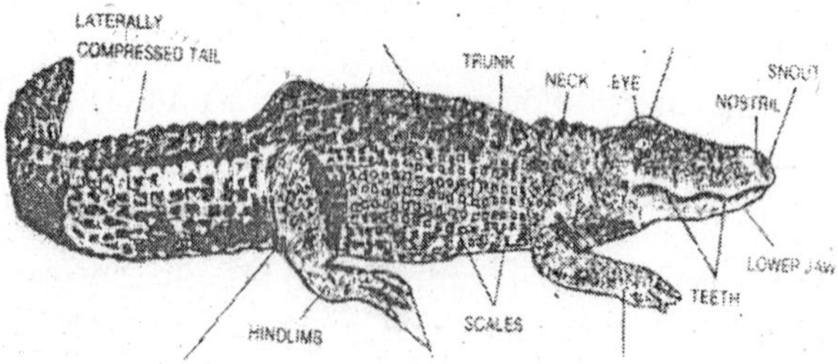

Alloecoela. Order of flatworms (Turbeilaria: Platyhelminthes); characterised by simple pharynx and diverticulated Intestine.

Allosteric Site. Activity of certain enzymes is regulated by modulator or regulator molecules which Interact with a part of molecule different from active site. This part of enzyme on combining with modulator brings a change in the tertiary structure of enzyme changing conformation of active site and thereby decreasing/Increasing (as in some cases) affinity between enzyme and the substance. Allosteric enzymes operate at key-points in metabolic reactions/pathways and modulation of their activity has significant role in regulating such pathways.

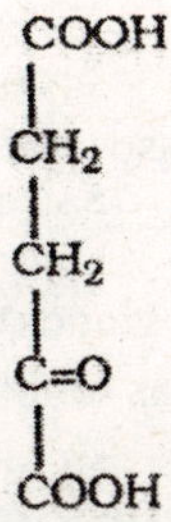

Allotriognathi. Equivalent name for order Lampridiformes of teleost fishes (Pisces).

Alpaca. South American domesticated hoofed mammal *Lama pacos* (Camelidre) akin to ca.'11el or llama; bred for its wool and also used as beast of burden.

Alopecia. Pathological disorder caused by deficiency of panthollienic acid; characterised by falling of hairs.

Alpha-Cells. (α-cells) : group of cells' in the islets of Langerhans (constitute 20% of these structures) which contain granules that stain red with Mallory stain. Secrete glucagon which is glycogenolytic hormone and converts glycogen to glucose.

Alpha-Keto Glutric Acid. (α-Keto Gluteric Acid): organic compound formed during Kreb's cycle by dehydrogenation of isocitric acid and converts into succinic acid by its own dehydrogenation. Also formed in tissue by glutamic acid during transamination. (Glutamic acid + Pyruvic acid = aketo gluteric acid + Alanine).

Alpha-motoneurons. Neurons exiting the spinal cord and synapsing with extrafusal fibres; initiate muscle' contraction.

Alternation of Generation. Metagenesis: alternative succession of sexual and asexual generation in life-cycle of an organism.

Altricial. Young one born or hatched· immature and therefore helpless and in need of extended parental care.

Alveolar Ducts. Duets arising from each respiratlry bronchiole in mammals and terminate 1n air sac whose walls are outfolded to formalveoll.

Alveolar Glands. Multicellular glands (acinous glands) have flask-shaped bodies and their necks serve as ducts *e.g.* mucous and poison gland of forgs and toads. Simple but branched alveolar glands have few saccules each leading into main duct *e.g.* glands along eye lids and gland associated with hair. Compound alveolar glands consist of a number of subunits called lobules and each lobule is formed of several simple sacs whose tiny ductules open into common ducts later connecting with main duet of the gland *e.g.* mammary glands. *Membrane*: periodontal membrane: tooth in mammals placed in bony socket is called alveolus; part of tooth inside the bony socket is called root witich is separated from bone of eveolus by collagenous periosteum of Jaw bone which is referred to as alveolar membrane.

Alveoli. Small pits/cavities *e.g.* tiny air spaces In mammalian lung; secreting portion of analveolar gland: sockets of teeth; minute chambers present on inner side of lungs of frog formed by the divlsion of it through a network of partitions; terminations of respiratory branchioles in mammals (number being 750 millions in both the lungs). Alveoli of lungs arc surround led by a capillary network of blood vessels in which exchange of gases (oxygen and carbon dioxide) takes place. Alveoli of glands are also supplied by blood capillaries for providing necessary substances to be required for the secretion.

Ambient. Surrounding/prevailing conditions *e.g.* ambient temperature, ambient pressure etc., to which the living organisms are subjected.

Amblyopsiformes. Equivalent and old term for fishes-order Percopsiformes (Actinopterygii: pisces).

Amennorrhea. Absence of menstrual periods in human female; two types - primary. if menstrual bleeding has never occurred and secondary if there is cessation of cycles in a woman with previously normal periods. Primary amenorrhea may be due to abnormality in the development of reproductive organs while secondary amenorrhea is caused by pregnancy or may be due to physical. physiological. emotional. nutritionai and cenvironment changes as well as due to overactomy.

Ametabola. Without metamorphosis: insects that are primarily wmgless and do not undergo any metamorphosis.

Aminopeptidase. Exopeptidase that attacks terminal bond at free amino end of peptide chain liberating amino acids and dipeptides; found to contain Mg++ or Mn.

Amitosis. Direct nuclear division during which nucleus constricts and separates into two portions while cell is in interphase without formation of condensed chromosomes. asters. spindles

Ammonotelic. Ability to excrete ammonia. Large quantities of water would be required to dissolve and carry off ammonia (300-500 ml for one gm of nitrogen as anuttonia); fishes (te1eosts) have this opportunity so they excrete their urine as a:.-nmonium hydroxide.

Amniocentesis. Removal of fluid from pregnant woman in order to study its composition and/or culture.

Amnion. Innermost extra embryonic membrane enclosing embryo in reptiles, birds and mammals; formed of extra embryonic ectoderm and extra embryonic somatic mesoderm, which corne out from the head and tail and end of embryo and grow laterally. dorsally except ventrally forming a pocket-like structure that later acquires amniotic fluid in its cavity.

Amniota. Group of vertebrates like reptiles, birds and mammals. in whose development extra embryonic membranes *viz.* aminon, alla!ltois. chorion and yold sac are formed.

Amniotic Cavity. Space enclosed by growth of ar:mion (see amnion) towards lateral and dorsal sides of embryo; fined with amniotic fluid (see amniotic fluid).

Amniotic Fluid. Fluid like substance filled Inside amniotic cavity which provides optimum moisture to developing embryo and also protects against shorks (external) which may occur due to displacement of egg.

Amoeba. Univellular fa cellular animal found in water (both fresh and marine). Irregular body shape. body covered with plasmalemma; nucleus single and bun-shaped, locomotion by pseudopodia (see amoeboid movement), reproduction by binary fision and multiple fission. Animal is considered to be immortal except violent death. Parasitic amoebae are called entamoebae.

Amoeboid. Cell movements resembling like those found in *fmoeba. Amorphous*; having no definite fonn or shape - amoeba. *Amphiaster*: type of spicule as seen in some sp01l ges. *Amphiaster*; figure produced by two asters and spindle in cytoplasm in the division of nucleus during mitosis.

Amphibia. Class of first land vertebrates possessing gills at least in their larval condition and lungs in adult stage. Skin poorly cornified and used as respiratory organ. Poikilithermal and undergo hibernation during winter and aestivation during summer. Heart three-chambered. Return to water for breeding-·· frogs. toads. newts and salamanders.

Amphibious. Organism capable of leading life both in water and on land-crocodile. frog and snail.

Amphibolic. Having ability to turn forward as well as backward outer toe of some birds.

Amphiblastula. Blastula like sponge laiva with large celled vegetal pole (megameres) and small celled ciliated animal pole (micromeres); free simming and changes to gastrula.

Amphicoela. Small suborder of Anura (Amphibia) that includes amphlcoeJous vertebrae.

Amphicoelous. Having concavities on both the sides *e.g.* 8th vertebra of frog. the centrum of which is concave anteriorly posteriorly to articulate with convexities of 7th and 9th vertebra res pectively.

Amphicyon. Tertiary fossil carnivore mammal that was about 2,5 metres long and looked like bear but had backbone like that of a cat which enabled the animal t.o leap like cat; modern bears are said to be its descendents.

Amphicyonida. Family of fossil giant predatary carnivore manlmals.

Anoguduokiud. Organism having diploid set of chromosomes from each parent.

Amphilinidea. Order of Cestodarta (Platyhelminthes); tapeworms having protrusible proboscis. anterior frontal glands and no holdfast organs : parasitic on sturgeons and other fishes --*AmphUina* species.

Amphimexis. Union of nuclei of ovum sperm to fom a zygote. AMPHINEURA : class of Mollusca. havinl! two ventral and two lateral neIVe cords. further divided into two subclasses-Aplacophora and Polyplacophora.

Amphiplatyan. Flat on both the sides-vertebrae of mammals where centra and pleurocentra both have more or less flat surfaces.

Amphipoda. Order of Crustacea that includes small sessile-eyed .animals. flattened from side to side and having legs adapt.ed for jumping and swimming: mostly marine. some Bve in fresh water and few are terrestrial - sand hoppers.

Amphisbaena. Fabulous two-headed snake (Ophidia::Reptilia): snake-like lizards of tropical America its rounded tail looks like another head at anal end.

Amphitoky. Parthenogenetic reproduction of both males and females.

Amphuma. Unique amphibian having snake-like body about one metre long and very small legs; found in South-east USA, remains mostly in burrows under water raising Its head above water only to breathe.

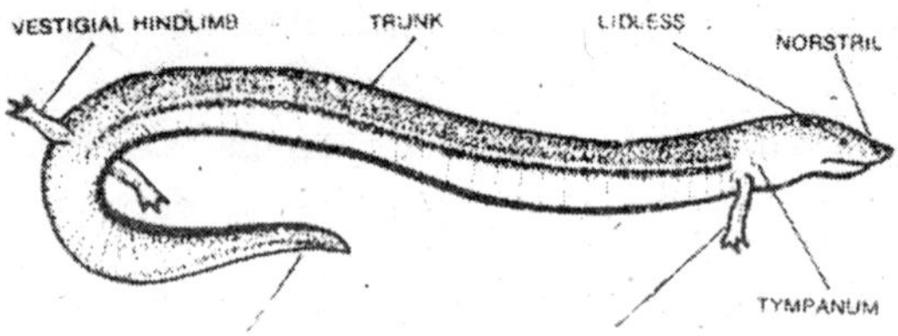

Amphotric. Chemical substance possessing qualities of both acid (capable of liberating H+ ions) and base (capable of accepting H+ ions) - amino acids.

Amplexary Pads. Sexual dimorphic character present in forelimb of male frog; formed in the first digit; grow and develop during breeding season under hormonal influence and serve to hold the female during amplexus (copulation).

Amplexus. Pairing in absence of copulatory organs as in frogs and toatis.

Amylase. Enzyme capable of hydrolysing polysaccharides to maltose (starch/polysaccharides - Dextrins, achrodextrins, maltriose and maltose); found in pancreatic juice of all the vertebrates and also digestive juices of other animals, in the blood and in tissues other than alimentary canal. A major enyyme of saliva in mammals that initiates digestion right from buccal cavity.

Anabolism. Constructive phase of metabolism in which cells build protoplasm from food materials *e.g.*, protein synthesis; comprises of endergonic reactions and requires energy to start.

Aanaconda. Large. non-poisonous snake (Ophidia Reptilia), of South America, that grows upto 10 metres in length.

Anadromous. Fishes that live in sea and go to fresh water for spawning -*Salmon* species .

Anaerobic. Able to live in absence of free oxygen.

Analogous. Parts similar in function but not necessarily alike in genetic relationship-wings of birds and insects. *Organs*: organs having only similarity in function but not In origin-Wings of bird and wings of insect. They are significant In organic evolution (Compare with honologous organs).

Analogy. Agreement of similarity in function between things otherwise different.

Anal Style. A pair of small structures attached to sternite of IX abdominal segment in most insects; characteristic features and identification marks of male cockroach.

Anamerentona. Old and obsolete name for insect order Protura (Apterygota: Hexapoda).

Anamnia. Veretbrate without extra embrynonic membrane; a vertebrate whose embryo Is not enclosed in sac like structures -fish and frog.

Anaphase. Stage during mitosis/meiosis/ when chromosomes move from equltorial plate to opposite ends of spindle.

Anapophyses. Additional mamillary processes attached with lumber vertebra of mammals and present between postzygapophses and transverse processes.

Anapsida. Group of primitive reptiles having no ·vacuities in the temporal region. 'They are represented today by turtles. tortoises and terrapins.

Anastomosis. Union of some body structure *e.g* , two or more blood vessels. nerves. other structures.

Anatomy. Science related with study of internal structures viz tissues, organs, systems etc. of animals/plants. Founded by Andreus Vesalius (1514-1564) who wrote the book. 'On The Structure of' the Human Body' in 1543.

Anchovy. Small. marine, herring-like fish (Pisces). having rich pungent smell.

Androconia. Wing (Lepidoptera: scales and connected scent glands of male butterfly Hexapoda).

Anemone. Common name for anth020an. *Metridium dianthus;* cylindrical creature with crown of tentacles arranged in a circlet in flower-like manner; sessile with usually brownish/ yellowish colour. Body cavity consists of six radial chambers between which arc six pair of mesenteries or primary septa: between them there are secondary and tertiary septa; free edges of septa are expanded into digestive filaments. near the base; these filaments carry long thread-like acontia carrying nematocysts. Dioceous animals, sexual reproduction by budding.

Aner. Male insect specially male of anant.

Aneurism. Localised dialation of an artery wall. The elastic walls are unstable and tend to balloon and if collagen fibres sheath breaks down artery then it tends to undergo ballooning.

Angina Pectorts. Substernal pain develops in chest when myocardium becomes ischemic (Also see myocardial ischemia) during exertion as a result of accumulation of p-factor In muscles.

Anglerfish. Group of marine fishes (feleostei Osteichthyes) having compressed body. rough skin, modified fin spine hanging erect above the snout with flap of skin on its top that acts as batt for small fishes around. found mostly at the bottom of Indian seas *Antennarius htspidus.*

Anguifauna. Assemblage (fauna) of snakes. ANGUIFORM : snake-shaped

Anguiliformes. Large order of actionopterygian fishes (Pisces) comprising true eels.

Anhingi. Family of aquatic birds (Pelecaniformes: Aves) good swimmers and divers, nest in colonies- anhinga or snake-bird.

ANI. Shiny black insect-eating bird (Aves) that lives in groups by constructing communal nests; found in America.

Animal. Organised living being having sensation and vohmtry motion. *Kingdom.* Entire animal fana that h'3.ve come into existence during last 2000 million years. *Pole*: in telolecithal eggs. The end where cytoplasm congregates.

Animalcule. Extremely small animal (microscopic, that cannot be seen with naked eye.

Animation. Animals-like activity; activities that are not totally natura! but stimulate the essential features of normal activities *e.g.,* living activities of frog dUring hibernation which resemble the normal life activities in essential features but not exactly similar to them.

Anion. Negatively charged ion. attached to anode or positive pole.

Anisochela. Chelate sponge spicute with dissimuar ends.

Anisodacyclons. (birds) Having unequal digits; three toes forwarn and one backward.

Anisogamy. Sexual fusion in which male and female gametes are of unequal size - fusion in malarial parasite *(Plasmodium)*

Anisoptera. Suborder of Odonata (Hexapoda); long slender insect with large eyes that meet mid-dorsally and wings held horizontally, strong 1liers. voracious feeders. prey upon small insects on wings - dragonflies.

Ankylosis. Union or knitting together of two or more bones or parts of bones.

Annovulatory Cycles. Menstrual cycles in woman during which no ovulation occurs.

Annulata. Old and equivalent name for phylum Annelida- the segmented worms.

ANOA. Wild ox looking like small bufallo, native to Celebes.

Anomura. Crustaacean suborder of Decapoda (Eucardia Malacostraca) - *Eupagurus* spp. Hippa

Anoplura. Order of blood sucking insects (Hexapoda), mostly ectopassites of vertebrates.

Anoxia. Lack of oxygen in tissue which diving animals *e.g.,* mammals and birds are subjected during submergence. The effect of it is first on central nervous system so the diving animals solve this problem by utilising oxygen stores in the body. oxygen is utilised only by brain and heart during this period and blood - flow to other parts/organs is reduced and these tissues adapt to anaerobic pathways.

Anseriformes. Order of aquatic birds (Aves) having desmognathous skull, flattened bill with horny rim round the edge and webbed-feet-ducks *(Anas* spp.) geese *(Anser* spp.) swans *(Cygnus* spp.) etc.

Ant. Small social insect (Formic1dree: Hymenoptera) that lives in colonies. characterised in having elbowed antennae and very narrow abdominal constrction - black ants are common household pests. while red ants are great nuisance in orchards. *Bear*. Common name of aardrark. the ant-eater found in Africa. *cow:* aphid (Aphididre: Homoptera) kept and tended by ants in their nests for the honey dew excreted by the aphid. *Eaters*: various mammals that feed· on ants; all having elongated snout and thin long sticky tongue - banded ant-eater. scaly ant-eater (pangolin), spiny ant-eater (edentate) etc.

Ant Hill. Very small hillock type structure. constructed by certain species of ants as their next.

Antennal Gland. Major organ of osmoregulation in crustaceans like cray-fish. Part of it, coelomosac. contains tubules which resemble in structure with vertebrate glomerullis; also exhibits structures that are physiologically analogous to proximal 'and distal tubules of the vertebrate kidney.

Anthozoa. Class of phylum Coelenterata that includes corals. sea-anemones etc.

Anthropoidea. Suborder of Primates (Mammalia), having plantigrade gait. eyes facing forward; nails instead of claws: diurnal - monkeys. apes and human beings.

Antheropomorphic. Scribing human characteristics to other organisms.

Antibiotics. Substance produced by micro-organism that inhibits the growth of another micro-organism. Widely used as drugs to combat bacterial disease *e.g.,* penicillin from Penicillium notatum is active against many staphylococcal infections and many gram-positive bacteria and sterptomycin obtained from *Streptomyces griseus* used for treatment of tuberculosis. Biosynthesis of antibiotics may be from amino- acids (penicilin), sugars (streptomycin), acetate or propionate (tetracyclin). Commercial production is done by large scale cultures on appropriate organisms through some simple antibiotics. They inhibit protein synthesis on 70 S ribosomes of prokaryotic organisms but not 80S ribosomes of eukaryotic organisms, Some antibiotics (cycloheximide) inhibit protein synthesis by 80 S ribosomes.

Antibody. Immunoglobulin: 4-chain protein molecule of specific amino-acid sequence: interacts only with the antigen that brought about its production or very similar to it. A protein produced in an animal body which neutralises or reacts with a foreign substance. Produced by lympth nodes.

Anticoagulant. Chemical substance that inhibits the tendency of thrombosis or blood clotting by antagonising the blood clotting factor, by destruction of key substance fibrinogen or by inhibiting the synthesis of factors, II, VII. IX and X *e.g.,* heparin.

Anticodon. Sequence of 3 nucleotides in t-RNA which pairs with specific codon in m-RNA. It is part of t-RNA responsible for giving identity to an amino - acid in cell for positioning it at proper place in the polypeptide chain.

Antimycin. Antibiotic isolated from *Streptomyces* strain; acts to black electron transport system by inhibiting transport of electrons from cytochrome-b to cytochrome-c.

Antitoxin. Chemical substance capable of neutralising other chemical substance which is a toxin-injection of atropin on exposure to insecticides.

Antlers. Pair of bony projections from skull of deer; unlike horns these aro often branched and are shed annually; mostly confined to males, except reindeer.

Aorta. Large artery that leaves heart; divides into arteries, arterioles and capillaries. In man, aorta has cross sectional area of about 4.0 cm^2 with cardiac output of 5.4 litres per minute, 90 ml of blood will pass through its lumen each second at a mean velocity of 22.5 em/second. As it subdivides, the radius of branches decreases but total cross sectional area increases.

Aortic Arch. One of the paired arteries which connects the ventral and dorsal aortae in the pharyngeal or gill region. *Body.* nodule on aortic arch containing chemoreceptors that sense the composition of blood; monitors changes in 02 and C02 in arterial blood *e.g.*, carotid body of mammals and birds, carotid labyrinth of amphibia. The chemorcceptors are innervated by branches of IX (glossopharyngeal} and X (vagus) cranial nerves.

Aoudad. Wild sheep (Artiodactyla: Mammalia) of North Africa.

APE. Large tailless monkey (Primate; Mammal), most nearly related to man.

Aphasmida. Nematodes lacking sensory organs near posterior end *-richtnella* (muscle worm).

Apiarian. Pertaining to beehive.

Apiarist. Person who keeps apiary .

Apiary. Place where bees are kept and reared.

Plate. Group of cells at anterior end of some trochophore larvae.

Apiculturist. Person who practises apiculture.

Apiologist. Entomologist who has specialised in study of bees, especially honeybees.

Apitoxin. Poison contained in bee's sting.

Apivorous. Feeding on bees.

Apne. Suspension or absence of breathing; brief periods of respiratory arrest. May be produced by injections of varatridine and nicotine into coronary and pulmonary circulation producing bradycardia and hypotension and due to stimulation of chemoreceptors.

Apneusis.. When all cranial nerves are cut and brain stem is transected above pons, regular breathing continues, however, an additional transection in inferior portion of pons results inspiratory neuron discharge continuously leading to a sustained contraction of inspiratory muscles and such arrest of respiration during inspiration is called apneusis.

Apocrine. Gland in which the products collect at the outer part of cells and are squeezed out with a little cytoplasm thus damaging a part of the cells sweat glands of mammals.

Apocrita. Suborder of Hymenoptera (Hexapoda) characterised by deep constriction between thorax and abdomen, these include ants, bees and wasps.

Apoda. (Gymnophiana) small order of legless amphibians: body wormiform, skin smooth with transverse ridges; eyes functionless, burrowIng habit, found in moht earth *Coecilia, Lchthyophis, Hypogeophts. Podemes*: plane of breakage in a crustacean appendage where autotomy occurs processes of Cixoskeleton enter the body forming the endoskeleton of insects because they form articulating sites for muscles.

Aponezyme. Protein portion of enzyme. It together with coenzyme/ cofactor forms the functional holoenzyme.

Apophyses. Various kinds of projections arising from a vertebra *Le.,* zygapophyses, tranverse processes etc.

Apopyle. Pore leading from flagellated chamber into central cavity or spongocoel of sponges; associated with sycon type canal systems.

Appendage. Portion of body that projects and has a free end - fore and hind limbs.

Appendicular Skeleton. Skeletal elements of girdles and limbs collectively constitute apendicular appendage; includes pectoral and pelvic girdles.

Apposition Image. Mosaic vision: during intense light. rays of light emitted by some objects in front of ommatidia of compound eye (in insects etc.) penetrate through cornea. All the oblique rays of eye will strike with iris pigment and terminate there while straight rays of light reach upto focusing part of ommatidium and the image of only that part of the object is formed from which the straight rays passed thus separate images will be formed by each ommatidium. This is called apposition image and is magnified and more clear.

Apterygota. Insects that are primitively wingless and undergo no metamorphosis; these include springtails (Collembolal, telsontails (Protura), campodids (Diplura), silverfish (Thysanura) etc., but not bugs, fleas and lice which. are secondarily wingless due to their parasitic nature.

Apteylae. Area of bird's skin where feathers do not grow (compan with pterylae).

Arabellidae. Family of polychaete worms (Errantia: Annelida)

Aracjomod: any animal belonging to class Arachnida.

Arachida. Class of Arthropodous animals; body consists of cephalothora (head and thorax combined) and abdomen, formel bears two pairs of prehensile appendages (pincers) and foul pairs of walking legs– crab, mite. scorpion, spider etc.

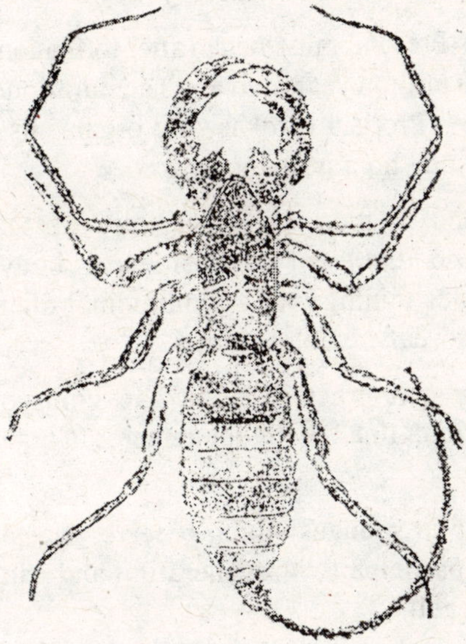

Aramidnee. Family of birds (Gruiformes: Aves) - limpkin. **ARANEAE** : order of Arachnida (Arthropoda) - spiders.

Araneida. Equivalent term for Araneae (Arachnida: Arthropoda) that includes spiders.

Arapaima. Pirarucu: pike-like, largest fresh water fish of the World (Osteoglossidce: Pisces). grows upto 2 metres in length and weighs upto 4 qUintals: found in South America.

Arapunga. Campanero: South American bell-bird (Aves): snow white in colour and its cry is like the sound of church-bell.

Arboreal. Pertaining to trees *e.g.*, tree living animals like monkeys etc.

Archaeocopida. Order of Cambrian crustaceans (Ostracoda) having only slight calcification of carapace.

Archaeornithes. Subclass of Aves (Chordata) that includes upper Jurassic birds which are all extinct by now.

Archaezoic. Geological period (Precambrain), began after formation of earth crust about 3,000 million years ago and had duration/ age of 2100 million years. Protista (prokaryotic organisms) evolved during this period; no fossil record.

Archenephros. Kidney found in most primitive vertebrates; consisting of one nephric tubule of each side per body segment and all the tubules joining into a longitudinal duct opening into cloaca. Also called holonephros.

Archenteron. Primitive digestive tract of metazoan embryo formed during gastrulation; lined with megameres (large yolky cells).

Archeocytes. Amoeboid cells in sponges which receive, digest and transport food; also participate in regeneration and can form any type of body cell.

Archeptrygium. Fish fin with single segmented axial rod from sides of which numerous secondary segmented radials arise; found in *Neoceratodus* species (lung-fish).

Archer. Fish small fresh water fish (Perciforms: Pisces) found in muddy coastal waters of India and South-east Asia, shoots its prey (Insects) with droplets of water.

Archiannelida. Small group of primitive marine annelids; body elongated narrow. internal segmentation well marked but not external, parapodia wanting and sexes united- Dinophilidce, Nerillidce and Protodrilidce.

Archousauria. Subclass of reptiles comprising six orders

Thecodontia, Pseudosuchia. Ornithschia, Saurischia, Pterosauria and Crocodilia; except a few crocodilians, rest are all extinct.

Archulus. Arc- shaped vein in wings of certain insects (Hexapoda) forming cross-connection between other veins of the wing.

Area Centralis. Small depression in the posterior side of retina (where there are more cones than rods) and its centre is exclusive region of cones called fovea centralis: in primates area centralis contains yellow pigment thus called macula lutea.

Areola. Distal extent of iris; closed cells of forewing of some Lepidopterans (group of butterflies etc.); longtitudirial strip of cray-fish. Carapace posterior to celVical groove: pigmented area around nipples of breast.

Argasid. Family of Acarina (Arthropoda) that includes soft-ticks.

Argonaut. Marine mollusc, closely related to octopus but having a thin white spiral shell; inhabits warm waters.

Aristotle's Lantern. Specialised calcareous structure surrounding mouth of sea-urchins consisting of teeth-like structures and helping in ingestion and chewing of food.

Armadillo. Nocturnal burrowing mammal, *Dasypus* spp. (Edentata); only animal haVing bony plates in the skin arranged in rings round the body; snout muzzle-like; tongue long and sticky for catching insects; found in America.

Arrector Pilli. Small bundle of unstriped muscle fibres attached to hair whose contraction brings about hair almost vertically erect to the surface of skin.

Arrhenotocia. Mode of parthenogenetic reproduction whereby only males are produced- queen honeybee.

Arryhthmia. In healthy young it:ldividuals breathing at normal rate; the heart rate varies with phases of respiration and this is called arrhythmia. Sinus arrhythmia (regular variation) is normal phenomenon. If it is due to abnormal pacemaker.

then it may cause articular/ventricular arryhthmia which may lead to myocardial disturbances (Also see extrasystoles and myocardial ischemia) .

Arterioles. Branches of an artery with lesser diameter and divisible into capillaries.

Arteriosclerosis. Disease of aging. characterised by infiltration of cholesterol into certain leisions of arterial wall; distorting them and making them rigid. It predisposes to myocardial infractions, cerebral thrombosis. The patients with high circulating lipopropteins are generally prone to this disease.

Artery. Division of aorta; branch of main aortic arch which divides into arteioles; blood vessel having flow of blood from anterior forward to posterior backward direction.

Arthropoda. Triploblastic metazoa having segmented body and jointed appendages includes crustaceans. myriapods. and arachnids. Most successful animals with over one million species and constitute about 78% of known species, of animals.

Artiodactyla. Mammals with even toed hooves like cow bufalloe , sheep. goat. camel. antelope etc .. two to four digits with hooves; elaborate ruminating stomach; having true homs. Beneficial animals for dairy. agriculture. wool industry. meat etc.

Arvicola. Water-rat: blunt-nosed. short-eared. mouse-like rodent (Rodentia: Mammalia).

Arytenoid. Pertaining to a pair of laryngeal cartilages lying posterior to thyroid cartilages in mammals; anterior to cricoid cartilages in other vertebrates.

Aschelminthes. Heterogenous phylum of worm-like animals; triploblastic with pseudocoel and alimentary canal; (1800 species) - *Ascaris. Enterobius* etc.

Ascon Type Canal System. Simple type of canal system found in sponge *Leucosolenia* species where diploblastic body wall is perforated at regular intervals and flagellate cells (choanocytes) line the entire spongocoel.

Ascorbic Acid. Vitamin-C; water soluble substance; necessary for hydroxylation of lysine in collagen synthesis; found in citrus fruits. leafy green vegetables etc.; daily requirement for infants is 35 mg. children 35 mg and adults 45 mg (both for men and women); deficiency causes Scuivy.

Asphyxia. Condition produced by oelusion of the airway *e.g* .. during strangling; pronounced stimulation of respiration with voilent efforts; blood pressure and heart rate rise sharply. blood pH drops. nor-epinephrine secretion is greater than epinephrine and eventually respiratory efforts cease. blood pressure drops and heart rate slows. Death occurs due to cardiac arrest if artificial respiration is not started within 4-5 minutes.

Aspidochirot. A order of Aspidochirotacea. holothurioid echinodernms having respiratory trees and dorsal tube feet modified into tactile warts - *Actinopyga* spp., *Stichopus* spp.

Aspidogastrea. (Aspidocotylea) : order of Trematoda (Platy helminthes) endoparasitic worms with well developed ventral holdfasts.

Aspidogastrid. Family of Aspidogastrea (Trematoda: Platy helminthes) endoparastic on molluscs - *Aspidogaster conchicola* etc.

Ass. Donkey: quadruped mammal related to horse.

Assimilation. Changing of absorbed food and other materials from intestine into cytoplasm.

Assortative Mating. Mating where mates are chosen as opposed to randomly selected mating.

Aster. Star-like figure formed during mitosis as well as meiosis and composed of centrosomes and lines radiating from it.

Astigmatism. Defocusing phenomenon in eye caused by deformation of cornea. In this defect oblique lines do not focus correctly since impinging of light rays are deflected from what would be considered a normal course.

Asymmetrical. Animal plant whose two sides are not alike: without any symmetry.

Ataxia. Inco-ordination due to errors in rate. range. force and direction of movement; caused due to leisions in cerebellum. There is no abnormality at rest.

Atlas. First vertebra in all the vertebrates. always modified to articulate with skull/head anteriorly and vertebral column posteriorly; first cervical vertebra in amniotes (reptiles. birds and mammals) upon which skull rests.

Atoll. Horse-shoe-shaped or ring-shaped island of a helt of coral reef surrounding a central lagoon.

Atrial Cavity. Cavity in protochordates and some caudate larvae (*e.g.* tadpole) through which water passes after moving out of gill slits.

Atrichous Isorhizas. *Steroline glutinants*: oval in shape with small thread bearing spines; help in locomotion of certain animals like Hydra by adhering with the substratum through hypos tome.

Atrium. Chamber/cavity; may refer to atrium of heart; specialised outer cavity of *Amphioxus* (see auricle).

Auerbach's Plexus. Nerve plexus found between outer longitudinal and middle circular muscles of alimentary canal: supplied by sympathetic and parasympathetic system and controls muscular contractions of alimentary canal.

Auricle. Ear-shaped part/organ or lobe-like appendage. The number of auricle is one In fishes; two in amphibia, reptiles. birds and mammals. These are placed at the top of ventricle; have muscular walls and receive blood from without the heart (*i.e.*, right auricle from sinus venous and left from lungs.)

Auriculaaia. Larva of holothurioid echinoderms without oiliated bands.

Autonomic. Independent. self-governing. The term usually refers to a subdivision of nervous system governing involuntary activities (consisting of sympathetic and parasympathetic nervous systems).

Autonoic Nervous System. System of ganglia and nerves controlling involuntary actions of body *e.g.*, actions of ductless glands, viscera, -olood vessels and all organs containing involuntary muscle. It communicates with central nervous system by way of roots of spinal nerves (see sympathetic ami parasympathetic systems).

Autophagous. Self-devouring, bird capable of feeding itself from moment of hatching.

Autophagy. Process whereby aging cells are destroyed by own lysosomes.

Autosome. Any chromosome except sex chromosomes - man has 22 pairs of autosomes.

Autostylic Jaw. Type of jaw suspension when lower jaw is articulated with the quadrate bone *e.g.*, in frogs. lizards and birds.

Autotrophic Nutrition. Process of nutrition in which an organism manufactures ,its own food *e.g.*, nutrition found in *Euglena* were animal synthesizes carbohydrates with the help of its chloroplast frpm C02 and water in presence of sunlight (as seen in the plants.

Auto Type (heutotype). Specimen identified by original author as illustration/ot" his species.

Aves. Group of warm blooded animals to which birds belong; feathered bipeds capable of aerial life; have high temperature metabolism and normal body temperature ranges from 40-50'C (often 60'C); heart four chambered; oviparous and amniotes.

Avian. Pertaining to birds.

Aviary. Large cage or building for rearing and maintaining birds.

Aviculture. Bird-rearing.

Avifaura. Assemblage of birds found in a region.

Axial Skeletion. Skeleton supporting body axis; skull and vertebral column constitute this skeleton.

Axis. Second cervical vertebra; a line along anteroposterior axis of body around which parts are symmetrically arranged.

Axolotal Laryva. Pedogenic larva of any of several species of salamander *Ambystoma* spp.. found in Mexico and Southwest United States, having external gills, tail, limbs etc. and capable . of reproducing sexually without undergoing metamorphosis.

Axon. Fibre of neive cell/neuron conducting neive impulses away from cell body or cyton of which it is a part: elongated cylindrical process of neuron along which-actiop. potentials are conducted: myelinated or non-myelinated both. (Le., with or without lipoprotein sheath).

Axoneme. Complex of mecrotubules and associated structures found in flagellar or ciliary shaft; consists of nine doublets and two central microtubules; enclosed in evaginated extension of cell membrane. Microtubules are substructures of globular proteins (40A) called tubulln. In central pair these molecules are organised into about thirteen longitudinal protofllaments Fer tubule. Outer nine doublets have a similar structure of globular proteins organised into protofllaments and are assembled into one complete tubule-A and attached to an incomplete tubule-B. Tubule-A has two arms pointing towards tubule-B of next doublet clockwise. The arms which repeat at inteivals along tubule-A consists of protein dyenin which exhibits activated Arpase activity. Peripheral doublet also bears series of radial links.

Azygos. Unpaired anatomical structure e.g., azygos lobe of lung, azygos vein of mammals.

B

Back Bone. Vertebral column, composed of bone.

Back-cross. Mating of a hybrtd to either of its parents.

Bacteriocyte. Modified fat cell that contains symbiotic bacterta 'or bactertum-shaped rods; found in certain insects.

Badger. Burrowing, nocturnal, hibernating, quadruped mammal, *Meles vulgaris* having plantigrade gait, coarse dark coloured fur with white blaze on forehead.

Balanced Carrier. Individual whose karyotype appears abnormal but who nevertheless has a normal chromosomal complement. Usually the result of translocation.

Baluchithere. Gigantic tertiary (25 million' years ago) fossil rahinocerus, *Baluchitherium* species (Pertssodactyla), largest land-dwelling animal, found in Baluchistan. shoulder height 6 metres, no horns, herbivorous; with gradual disappearance of forests, these beasts died of starvation.

Bandicoot. Pig-rat: longest species of rat found in Indian subcontinent, *Nesocia bandtcota* (Rodentia: Mammalia): Insectiorous and herbiorous marsupial, *Permeles* species found in Australia.

Banteng. Wild ax (Artiodactyla: Mammalia) found in East India and Malaysia.

Barbels. Slender fringed bipectinate tactile appendages surrounding mouth of lemprey (*Agnatha*) : two pairs of black -coloured sensory organs arising from rostrum and maxillae of certain carp and catfishes of freshwaters, such as *Labeo* spp.

(*Cyprinidre*) : Large, freshwater fish of Europe, *Barbus* species (Cypriniformes : Teleostel) having beard-like appendages hanging near its mouth.

Barbicel. Small, hook-bearing process on a barbule of a barb of a feather.

Barbs. Lateral cornified branches from main rachis of a feather that divide into barbules.

Barbules. Processes from barbs of a feather which carry barbicels or hooklets.

Barnacle. Aquatic shell-fish (Cirripedia ; Crustaceae) which adheres to rocks and ships' bottom.

Baroreceptors. Sense organs/receptors in the body which respond to change in pressure in the body. They are distributed widely in arterial system *e.g.*, carotid sinus, aortic arch, subclavian, common carotid or pulmonary arteries. Increase in blood pressure would cause increase in discharge frequency from these since they are utmost sensitive to physiological range of blood pressure.

Barracuda. Snock : common name for over 20 species of edible fish. *Sphyraena* spp. (Perciformes: Pisces) found in South African warm seas.

Barr Body. In human female's egg cells, a dark staining portion visible under light microscope. It is actually one of the two X-chromosomes and is inactivated.

Barrier. Obstruction- physical, chemical or biological- that prevents migration of animals or gradual extension of their territories.

Barrier Reef. Reef separated from shore by a wide. deep channel.

Basal Body. Kinetosome: a microtubular structure from which a cilium or flagellum arises; considered homologous to centriole.

Basal Metabolic Rate. Energy output of an organism under standard resting conditions *(Le..* complete body and psychological rest). 12-18 hours after a meal and in an equable environmental temperature. Normal average sized man has BMR of 2000 koaljday. Body surface area, age. sex, race. emotional state, climate, body temperature, circulating levels of hormones affect the BMR.

Basilar Membrane. Delicate ribbon-like tissue bearing auditory hair cells in cochlea of mammalian ear.

Basket-Stars. Common name for small echinoderms (Gorgonocephalidae : Ophiuroidea) - *Astrophyton* species. *Gorgonocephalus* species etc.

Basophils. Granulocytes (WBCs) of blood constituting about 0.4% of WBCs. Cytoplasmic granules are large sized but fewer; nucleus is characteristically large sized and constricted into several lobes not joined by delicate threads. Non-phagocytic. These are believed to contain heparin and histamine but their role in balancing clotting and anticlotting remains imperfectly understood.

Bat. Mouse-like flying mammal (Chiroptera), no true wings but fore-and hind limbs connected with each other by means of extended skin of lateral sides of trunk; nocturnal, find their way in absolute darkness by using highly developed sonar or echo-apparatus - *Pteropns medias, Scotophilus kuhli* etc. *Vampire*: blood-sucking bat of Central and South America.

Bathyal. Ocean zone comprising upper 200 meters of water.

Beak. Pointed projection or protuberance of face; present in all birds as also in Umbo (bivalve molluscs), certain fishes (sawfish, pike etc.) and other animals (duck-bill platypus and turtles etc.)

Bear. Heavily built carnivore mammal having long shaggy hair, very short tail, broad feet with five toes and non-retractile hooked claes .- *Ursus torquatus, U arctos, Selenarctos thibetanus* etc.

Beaver. Large amphibion rodent (Rodentia: Mammalia) that grows upto one metre, with broad flat tail. hind toes webbed. lives in colonies, burrowing near banks; valued for its fur.

Beavery. Place where beavers are kept and bred.

Bed-Bug. Common name for insect *Cimex* spp., ectoparasite with piercing mouthparts, no wings; found in cracks and crevices; carrier for several diseases of man.

Beetle. Insect whose fore-wings are modified into two hard protective structures called elytra (Coleoptera: Hexapoda)

Behaviour. Relative responses/reactions of organism with respect to its environment.

Bell-bird. Campanero: common name' for number of birds (Passertfornes : Aves) that make tinkling nott' clear bell-like call.

Bell-Magendie Law. Dorsal root of spinal cord contains only sensory axons whereas the ventral root contains only moter axons.

Benthic. Organisms living on or in the bottom either of sea or fresh waters: from edges of water down to the greater depths.

Beri-Beri. Disease caused by deficiency of vitamin B} in the diet and characterised by disturbed functioning of nervous and muscular systems, weak cardiac muscles, constipation, indigestion etc., may often lead to parlysis / cardiac failure.

Beyrichicopina. Suborder of fossil ostracods (Paleocopa Crustacea).

Beyrichiid. Family of fossil ostracods (Beyrichacea Beyrichicopina)

Beta-keratin. Insoluable sulpher-rtch scleroprotien ; constituent of epidermis, horns. nails, feathers and tooth enamel. Beta refers to protein's secondary structure which is 1Il pleated sheets *(e.g .,*

silk produced by caterpillers).

Beta-Oxidation. Oxidation at beta-position, a pathway 'envolving removal of two carbon atoms at a time from carboxyl group (end) of molecule. Knoop proposed such oxidation for fatty acids. 13-oxidation of fatty acids results in the formation of two molecules of acetyl Coenzyme- A

Bicuspid Value. Mammalian heart the bilobed left auriculo-ventricular valve.

Bidder's Canal. Male amphibia (frog) some urinary tubules connect with testes and transport spenns into a duct called Bidder's canal that runs along medial edge of and outside the kidney and communicates with ureter.

Bilateral Symmetry. Arrangement of parts of an organism in such a way, that the right and left halves are mirror images of each other.

Bile. Fluid secreted by liver in vertebrates; contains about 97% water, 0.7% bile. salts, 0.2% bile pigments, 0.06% cholesterol, 0.7% inorganic salts and 0.1 % fats alongwith little alkaline phosphatases; makes food alkaline, prevents putrifaction and reduces surface tension over fat moelcules for making them emulsified.

Bilirubin. Formed by breakdown of haemoglobin; one of the bile pigments responsible for golden yellow colour of faeces. Its accumulation in blood, skin scleres and mucous membranes imparts these structure a golden yellow colour and such a condition is clinically known as Jaundice.

Bilvirdin. Green bile pigment formed by breakdown of haemoglobin. This is done by reticuloendothelial system. In man most of bilivirdin fonned is converted to bilirubin.

Binocular Vision. When central parts of visual fields in both the eyes coincide, anything in this purtion is seen as one object from both the two eyes and such a vision is called binocular vision. This type of vision is seen among human beings.

Binomial. An expression consisting of two tenns *(e.g.,* a and b) separated by *a*+ or *a*- sign. A binomial may be expanded by an exponent such as *(a* + b) . Expansions of binomial to a particular power has great value in determining probability that a certain combination of events will occur. - ***nomenlature*** : system of naming organisms (plants qr animals) suggested by Carolus A Linnaeus (1707-78) in 1735. Every organism described is given a generic and an specific name, followed by name of person who first described the specimen. Scientific name of green parrot is *Psutacula* kramer! (Scopoli), horned-owl Bubo bubo (Linnaeus) and mar.. *Homosapiens* Linnaeus.

Bio. Life

Biogenesis. Theory of origin of life according to which living things are produced only from living things (Compare with abiogenesis) .

Biogenetic Law. Recapitulation theory: principle that an animal repeats corresponding stages of its ancestors during its embryonic development or 'Ontogeny Repats Phylogeny'. Proposed by Von Hackel. .

Biological Father. Actual father: the person who has sired the off speing whether married or not with the woman concerned. *Membrane.* Plasma membrane. Generally 7.5 nm (75;\) thick and made up of protein and lipids, mainly phospholipids *e.g.,* phosphatidylcholine and phosphatidyl ethanolamine.

Biological Oxidation. Oxidation is combination of substance with O_2 or loss of hydrogen/electron. Biological Oxidation refers to such processes which are related to production of energy in cell. Catalysed by a particular protein enzyme.

Biologist. Scientist who has speciallsed in biology.

Biolog. Science that deals with study of living things; includes zoology and botany which are concerned with study of animals and plants respectively.

Bilouminiscence. Production of light, as in fishes, due to chemical reactions in living organisms.

Biome. Habitat zone *e.g.,* grassland or tundra resulting from interaction of climate, biota and substrate.

Biosynthesis. Synthesis of some organic molecule within living organism through a series of biochemical reactions each step being catalysed by a specific enzyme *e.g.,* synthesis of hormones, fats etc. in the body.

Biota. All organisms present in a particular blome.

Biotin. Member of vit4miI\ B group; catalyses CO_2 fixation in fatty acids' synthesis etc., deficiency causes dermatitis and gastrointestinal disorders; found in egg yolk. liver, tomatoes etc.

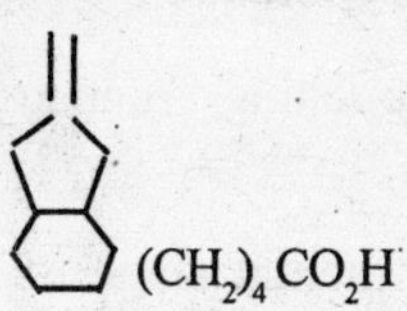

Biped (Bipedal). Animals having only two feet -,birds (Aves) etc. BIPINNARIA : larva of star-fishes with two ciliated bands resembling wings.

Bipolar Cells. Neuron with two axons emerging from opposite sides of the soma. One such class is found in retina of the eye where they serve to transmit signals from visual receptor cell to ganglionic cells of optic nerve.

Biradial Symmertry. Condition in which an animal has radially arranged parts that lie half on one side and half on other side of a median longitudinal plane. It is found in ctenophores.

Biramous. Crustacean appendagesd with protopodite, endopodite and exopodite; each of two branches is segmented and some branches may be absent.

Bivalent. In pachytene stage of meiosis, the four chromatids formed from two haomologous chromosomes are joined together at centromere; also called tetrad.

Bladder. Distended membranous sac for storage purpose. -, AIR : bladder that contains air. *Urinary:* receptacle for urine in animal body. *Gail*: Bladder storing bile. *Worm*: cysticercus; tapeworm larva found in voluntary muscles of secondary host (pig). Larva contains a small scolex, short neck and a bladder or vesicle ruled with fluid. Infective to man when improperly cooked measley pork is eaten. In intestine of man, larva develops into adult.

Blastomere. Cleavage cell anyone of the cells In an embryo from first cleavage division to the beginning of gastrulation.

Blastopore. Pore-like opening from archenteron or gastrular cavity to outside.

Blastula. Early developmental stage of an embryo. Usually a hollow mass of cells with a cavity inside which is called blastocoel.

Blauwbok. Small South African sahle antelope (Artiodactyla Mammalia) including various extinct species.

Blaze. White patch or horse's face running from forehead to nose.

Bleeding Time. Time when bleeding from some injured part ceases. Normally 2-6 minutes in most of the cases except under purpura when It exceeds from 10 to 20 minutes.

Blennioidei. Large suborder of small marine fishes (Perciformes:

Pisces) that live mainly in and around coral or rock-reefs.

Blenny. Small, slimy. spiny-finned acanthopterygian fish. *Blennius* spp. (Blenniidre : Pisces) found on rocky seashore. tidal pools and estuaries.

Blepharoplast. Basal granule; basal granule associated with flagellum.

Blebock.. Small sturdy South African antelope (Artiodaetyla: Mammalia) with blazed forehead and short horns more or less parallel.

Blind-spot. Point in retina of eye at which axons of ganglionic layer converge .into optic nerve; contains rods and cones; incapable of producing impulses or forming image.

Blister Beetles. Spanish-flies: insects (Meloidre : Coleoptera) having soft cylindrical body. 15 to 25mm long with distinct neck; contain cantharidin oil which when applied to skin produces blisters; pests of vartous vegetables and ornamental - Mylabris spp. are polyphagous pests.

Blood. Fluid circulating in vascular system of man and other animals; liquid connective tissue consisting of 60% fluid portion called plasma and *400A!* cellular portion called corpuscles (RBes. WBCs and platelets). Serves to transport substances from one part of body to the other.

Blood-Brain Barrier. There is a biological barner between brain ECF and blood that permits water. carbondioxtde and oxygen to cross cerebral capillaries with ease and other substances slowly. This is due to H+ gradient between brain ECF and blood (pH of ECF is 7.33 and that of blood is 7.40). Develops since early stage. Maintains constancy of environment of neurons in CNS. It breaks down in areas of brain that are infected. irradiated or the sites of tumours thus has clinical application in diagnosing such diseases.

Blood Clotting. Phenomenon by which blood is immediately turned to clot-like structure where there is some injury and bleeding; facilitated by different clotting factors; brief events of this process are given on next pape. *Corpuscles.* Minute dise-like or spherical bodies constituting a large part of blood in vectcbrate animals; three major types -erythrocytes or red blood corpuseles (RBCs). leucocytes or white blood corpuscles (WBCs) and thrombocytes or plateles.

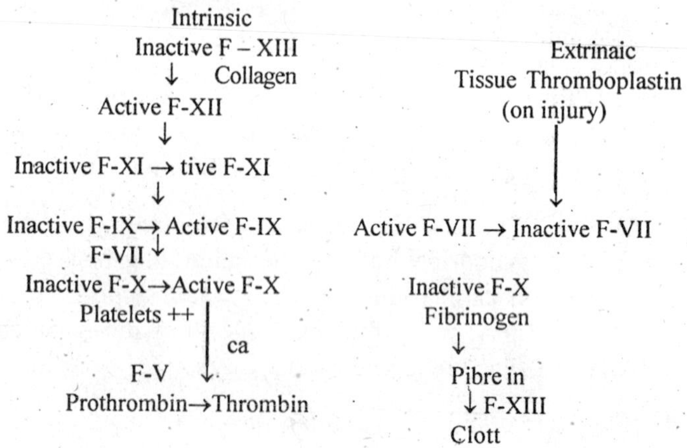

Blood Clotting

Blood Group. Discovered by Karl Landstiner (1868-1943) in 1900; classification of human beings on the basis of proteins found in RBCs (agglutinogens) and specific antibodies formed in serum (aggulutinins) most important of these are A and B (aggulutinogens) and a and (agglutinins). According to their presence/absence we recognise four types of blood groups A. B. AB and O. Other types known are AI' A_2• BO; MNS_s: P; Rh; Lutheran; Kell: Duffy; Lewis. Kidd. Diego; yt; Auberger; Dombrock: Colton: Sid; Scianna: Ian; Leavy; Evan; Gt; Heibel; Hey; Hov; Chido and Xg. A. B. AB. and a blood groups are concerned mainly with blood transfusion; Rh with pregnancy and MNSs are associated with patemity of the child.

BLOOD GROUPS OF CHILDREN AND PARENTS

Offspring	*One parent*	*Other Parent* *may be*	*cannot be*
A	*O.B*	A.AB	O.B
B	*a.A*	B.AB	O.A
AB	A	B.AB	O.A
B	A.AB	O.B	
AB	- A. B.AB	a	
a	a	A.B	AB
A	0.8		
8	O.A		

Plasma. dull yellowish fluid that carries blood cells (corpuscles) in human beings. it is 91 % water. 7% proteins and 2% digested food-stuffs. gases. salts etc. *Pressure*: force exerted by blood on walls of artery. Artelial pressure is the result of discharge of blood from left ventricle to aorta (which is already full). The systollic pressure is the result of pushing of blood into aorta due to contraction of left ventricle and is generally 120 mm of mercury. The diastolic pressure is produced by cardiac diastole and Is 80 mm of mercury. The systolic blood pressure Is calculated generally as (100 + age of the person and diastolic as 100 - age of the person mm of mercury). In females the blood pressure is 10- 20 mm of mercuryless than that of males of same age. Blood pressure is dropped by 20-30 mm of mercury during sleep. see hypo-and hypertensions. *Sucker*: animal that sucks blood of other animal - bedbug etc. *Sugar*: amount of glucose present in circulating blood. *Bloodless*: with out any blood.

BOA. Large tropical non-poisonous snake (Ophidia:Repitila) that. like python. kills its prey by compression or constriction; feeds mainly on rodents and squirrels - *Boa constrictor* is found in Brazil (South America).

Blue-back. Sockey: chief salmon. *Oncorhynchus nerka* (Pisces) of North Pacific ocean. *Bird*: common singingbird *Stalta stalls* (Sylviidre: Aves); having beautiful blue plumage.

Bottles. Blowflies: dipteran flies. *Calliphora erythrocephala. C.comttorta* (Calliphoridre:Tachinoidea) having metallic blue abdomen; lay eggs on carrion. meat etc., these hatch within a day and maggots feed on the flesh there by enlarging the injury. *Breast*: (blue-throat) small singing and perching bird. akin to nightingale (Passerifonnes: Aves). *Eye*: blue-cheeked Australian honey-eater (Aves); very fond of nectar. also eats small insects captured on the wing and makes, noisy sounds. *Fin*: tuna: large streamlined and fast swimming oceanic fishes about two metres long (some grow upto 4 metres) body dark blue dorsally and silvery beneath. *Fish*: shipjack: voracious predatoryjumping fish. *Pomatomus saltatrtx* (Serranidre: Percifonnes) weighing upto 10 kg. found in warm seas of the World, swims in large shoals and attacks all kinds of fishes around. *Hare*: mountainous hare: *Lapus ttmldus* (Lagomorpha: Manunalia) which changes its furcolour. to white in Winter; found in temperate regions from Scotland across northern Eurasia upto Japan. *Jay*: Indian neelkanth. *Coractas benghalensis* or American *Cyanocitta Cristata* bothPasseriformes (Aves) having vividblue and blackIshwings. *Shark*: large marine man-eating shark (cartilaginousfish) gmws beyond 12 metres in length and weighs as many tonnes. *Sheep*: mountain dwelling ungulate. *Pseudois nayar* (Artiodactyla: Mammalia) having spimlly twisted horns; found in northern hemisphere. *Whale*: large bluish-grey whale. *Balaenoptera musculus* (Cetacca:Eutheria) is the largest Living mammal. about 32 metres long and weighing upto 150 tonnes; lives inpolar seas during summer. move to warmer seas in Winter. feeds on tiny planktonic animals; newly born calves are over 7 metres long.

Boar. Male (uncastrated) swine Sus cristatus, S. serofa (Artiodactyle:

mammalia) about a metre long. with tufted tail another 25-30cm; body boawnish-black with few hair and movable snout with firm ·cartilaginous tip.

Body. Entire frame of man or any other higher animal: (slang) corpse. *Cavity:* cavity located between body wall and internal organs of an animal.

Bohr Effect. Interaction between H+ ions and haemoglobin haemoglobin affinity. An increase in H+ ions causes a reduction in haemoglobin-oxygen deficiency. Conversely oxygenation of haemoglobin results in release of protons from haemoglobin and decrease in pH of blood. Carbondioxide reacts with water to form carbonic acid and with NH_2 group to form carbamino compounds. Thus carbondioxide transport is facuitated through blood.

Bolus. Lump of masticated food with saliva.

Bombida. Family of social bees (Hymenoptera: Hexapoda) robust. black and yellow coloured Insects that nest In ground - *Bombus*

Bone. Hardest tissue In the body; connective tissue composed of collagen. calcium. phosphorous salts and ossein. Covered by periosteum. *Marrow.* Vascular connective tissue found at the ends of long bones. Active cellular marrow is called red bone marrow while that Inffitrated with fat is called yellow bone marrow. Red bone marrow consists of myeloblasts. haemocytoblasts and megakaryoblasts which form WBCs, RBCs and platelets respectively.

Bothridium. Muscular holdfast organ on head (scolex) of tapeworms. often with hooks.

Bottle-Neck Effect. Genetic drift where, as a result of rapid decrease In population size, the genetic pool Is drastically reduced thus gene frequencies of remaining genes are greatly Increased relative to original pretrauma population values.

Bowfin. Mudfish fresluvater fish. *Amta* callJa (Amudre : Pisces) of North America. with primitive features as possessed by fishes of Mesozoic era.

Bowman's Capsule. Cup-shaped end of kidney tubule that encloses glomerulus. Lined with squamous epithelium and helps in ultrafiltration.

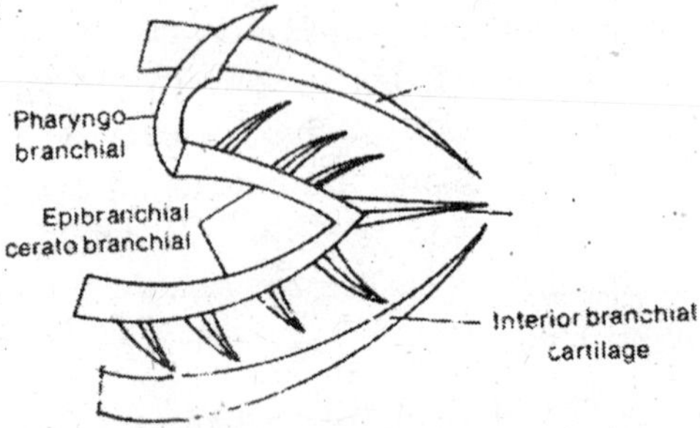

Brachium. Upper part of fore 11mb of a vertebrate. Anteriorly articulated with shouldt:r and posteriorly with antibrachium. Skeleton support is formed by humerus bone.

Brachypterous. (fishes) Having short fins: (Insects) having short wings that do not cover abdomen.

Bradycaria. Condition in which there is heart rate reduction from normal level.

Bradykinin. Hormone formed from precusor normally circulating in blood. It is very important vasadialator.

Brain. One of the divisions of central nervous system fonned by anterior part of neural. tube: consists of fore. mid and hind brains. Seat and controlling apparatus and co-ordinator of nervous activities of an individual.

Branchial Arches. Five pairs of embryonic visceral arches in the vertebrates are called branchial arches since they support the gills and pharynx allowing muscular movements of pharyngeal wall for a respiratory current. A typical branchial arch consists of 9 cartllagenous pieces. a lower median basibranchial to which are attached on each side a hypobranchial. a ceratobranchial. an epibranchial and a pharyngobranchial cartilages. The ceratobranchials support gill rays.

Bream. Gregarious fresh water fish. *Abram(s* species (Pisces) having high-arched back.

Breast. Human lactating apparatus arising as an invagination from surface epithelium that dips down into underlying connective tissue as a solid column; these are gradually hollowed out LO become ducts of mammary glands. Bearing a nipple at which the various ducts open and surrounding it is areola. Rudimentary at birth. little further development occurs until puberty and fully developed after delivery. Undergoes further proliferative change with recurrence to each menstrual cycle. Between each menstrual period there is pyperaemia (increase of areolar stroma} and possibly formation of new alveoli; these changes are transitional. During first half of pregnancy there is further appearance of many alveoli which form lobules but no milk formation. During second half the epithelial cells swell and gradually secretory activity increases.

Breathing. Physical process affected either by osmosis (aquatic animals) and diffusion (terrestrial animals) through which an exchange of oxygen and carbondioxtde is done between body and environment of an individual. Also called external respiration.

Brisingidat. Family of Forcipuloitida (Echinodermata) having small circular well defined disc with as. many as 44 arms; found only in deep-waters-Brtstnga species. Odinia species.

Bristle-tall. Small primitive apterygote insect (Thysanura: Hexapoda) with spindle-shape body covered with scales; antemare long and filiform. cerci long and multisegmented.

Bristle-Worms. Group of marine worms which have number of prominent bristles - Lugworms. ragworms.

Bronc Hole. Branches of trachea inside tissue of lung in some vertebrates *e.g.* mammals. They may/may not divide into primary. secondary. tertiary bronchioles and respiratory branchioles. All these branches except respiratory bronchioles carry cartilagenous rings like trachea and bronchi.

Bronchous. Branch of trachea in most of vertebrates. Each enters lung of its side and divides into bronchioles. etc. Bears cartilage no us rings protect the collapsing of its walls.

Bronothere. Thunder beast (Perissodactyla); Oligocene (Tertiary) fossil mammal shoulder height 1.5 to 2 metres; bulky body with two blunt horns on its nose; lived in forests of North America.

Drontotherioidea. Monofamrual (Brontotheriidre) super- family of Hippomorpha. comprising of extinct perissodactyl mammals.

Brown Fat. Fat found in infants but not in adult of human beings. It has a high metabolic rate and produces heat that helps in thermogulation. It is located between and around the scapulas.

Brownian Movement. Movement of cytoplasm and/or karayoplasm due to intracellular molecular collision specially seen in *Amoeba.*

Bruits. Abnormal sounds heard in various parts of vascular system; major cause is supposed to be the-disease of valves of heart.

Brush-Border. Free epithelium or epithelial cell surface bearing numerous. small microvilli *e.g.*, epithelium of proximal urinary tubule.

Bruxism. Teeth grinding during sleep in some individual; also associated with dreaming.

Bubo. Inflammatoty swelling of glands especially in armpit or gram. BUCCAL : mouth or oral cavity of an animal.

Bud. Developing lateral branch of an organism *e.g., Hydra:* term also used for yeung organism produced by budding.

Budgerigar. Zebra parakeet. *Melopstttacus undulatus* (Psittacifonnes: Aves). native to Australia; often kept as pet.

Buffers. Chemical system that establishes concentration of a substances; acid-base system serve as pH buffers preventing large changes in hydrogen ton concentration.

Bulbourethral Glands. Cowper's glands: paired glands surrounding urethra and associated with urinogenltal organs in mammals. secrete a mucous-like substance that comes out prior to ejaculation. It lubricates vagina, neutraltses acidity of vagina, helps in fertilisation by dissolving certain cells around the ovum.

Bulbus Arterious. Enlargement (bulb like) at the base of ventral aorta found chiefly In bony fishes.

Bulbus Arterious. Enlargement (bulb like) at the base of ventral aorta found chiefly in bony fishes.

Bull. Uncastrated male of ox: male of elephant, waln1s. whale etc. - *Calf*: male calf.

Bulla. Hollow bony growth such as bursa of a joint.

Bulldog. Large-headed, smooth-haired breed of dogs, having great courage: once used for baiting bulls.

Bullfinch. Strong-beaked. beautiful plumage, song-bird, *Pyrrhula pyrrhula* (Aves).

Bullfrong. Very large frog of *Rana* species (Phaneroglossa: Anura) with loud bellowing note.

Bullock. Castrated (sterne} bull.

Bungarotoxin. Group of neurotoxins Isolated from venom of snake krait *(BWlgarus* species): binds selectively and irreversibly to acetylcholine receptors.

Bunner's Glands. Small coiled acinotubular glands In the duodenum: secrete a thick alkaline mucous which probably helps to protect the duodenal wall from gastric add. These open In krypts of Loiberkuhn.

Bunodont Teeth. When cusps in molar teeth remain distinct from one another the tooth is called bunodont-man, monkey etc.

Bursa Entiana. Small sac-like structure present between pyloric stcmach and intestine in cartilagenous fishes like. Scouodon spp.

Bursa FAbricii. Thick glandullar pouch of lymphatic tissue lying just above cloaca in case of many birds pigeon: serves to fight infections of cloaca.

Bursicon. Neutrosecretory cells in brain and nerve cord of insects produce a harmone that influences certain aspects of cuticle development, including process of tanning which is completed several hours after each moult; small molecular protein with molecular weight of about 40.000.

Buch-baby. Small African, nocturnal. tree-climbing animal. *Galago* species (Primates: Mammalia); long-eared, large-eyed and long-tailed.

Bushibucb. Small African antelope. *Tragelophus scrtptus* and allied species inhabiting forests and swamps.

Bushpig. Red river hog; sto~t-bodie. Afncan pig with reddish-dark-brown body, big-head and tufted-ears-; feeds on fruits and roots.

Bustard. Group of large birds *Otis* spp. (Gruiformes: Aves) Stout-long legged with strong toes; swift runner that can also fly; found usually in marshy places.

Butcher Bird. Shrike. *Lanius* supp. (PasserifOnnes: Aves) having strong-hooked bill. predaceous on insects, mice and small birds.

Butterfish. Gunnel: eel-like fish *Pholis gunnellus* (Pisces) of North Atlantic, about 15 cm. long, brownish-green with dark markings.

Butterfly. Diurnal Lepidopterous insect (Hexapoda), body and wings covered with scales. antennre clubbed and wings usually held vertically above the body when at rest.

Buzzard. Any hawk, *Buteo* spp. (Falconiformes: Aves), large bird of prey, desplred by falconers.

C

Cabezon. Largeh eaded. marine, spiny-fish. *Seorpaentehthys mannol'atus* (Pisces), about 75 cm long, weighing' 11 to 12 kg and' good-for-nothing.

Caecum. Cavity open at one end *e.g.,* pouch that is beginning of large mtestlne.

Cahalot. Kind of sperm whale *Physeter* species (Cetacea: Mammalia) that grows upto 20 metres in length. has large head-cavity full of sperma.

Cake Urchin. Sea biscuit: flatterned form of sea-urchin. *Echtnus* spp. (Echinoidea: Echinodermata, shell about 10 em. across is purple to black in colour; found in warmer seas of Pacific ocean.

Cacomistile. Ringtail. racoon-fox or cat-squirrel small, furry, omnivorus mammal, *Bassartseus ustatus* characterised by bushy black and white ringed tall; found in South-western USA and Mexico; valued for its fur (California mink).

Cainotheioidea. Monofamiliai superfamily of fossil Artlodactyl mammals that included rabbit-sized tylopod ruminants.

Calanoida. Order of Copepoda {(Extomostra: Crustacea) that includes larger and more abundant pelagic species.

Calathiform. Cup-shaped or hemispherical

Calcaneum. Bony projection of metatarsus in birds (Aves): heel bone in human beings.

Canary. Small song-bird (finch or serin). *Serinus* canartus

(Aves) native of Canary Islands; bright-yellow to brownish-yellow In colour. feeds on seeds and grain; popular as cage-birds.

Canida. Family of camirorous digitigrade mammals. having four toes with claws in each foot - wild and domestic dogs (Cants spp.) and their allied.

Canine. Conical tooth: one located between lateral incisor and first premolar as in many mammals and human beings.

Capillary. Branches of metarteriole of arteiole opening surrounded by minute smooth muscles called precapillary sphincters. Connects metarteriole with a vennule.

Capitulum. Head-like rounded structure present with the ribs of mammals and birds and articulated with centrum of thoracic vertebra.

Caprimulgiformes. Order of crepuscular and nocturnal birds

Captaculum. Tentacle-like food capturing structure growing out from the head of tooth-shell mollusc.

Capuchia Monkey. Small gregarious monkey of South America *Cebus* species (Platyrrhina: Mammalia) with brownish-black hair. cowl at back of head and long prehensile tail.

Copybara. Water pig: largest tailless rodent *Hydrochrerus capybara* (Rodentia: Mammalia) grows upto 1.5 metres. inhabiting rivers of tropical South America.

Carapace. Hard dorsal shell of turtles and many crustaceans. CARBOHYDRASE: enzyme promoting break down or synthesis of a carbohydrate.

Carbohydrate. Aldehyde or ketone derivative of alcohol consisting of C. Hand 0. Utilised by animals primarily for storage and supply of chemiccl energy. Most important of these are sugars and starches.

Carboxyamino Haemoglobin. Haemoglobin when deoxygenated

forms carbamine haemoglobin by combining with carbondioxide and thus the transport of carbondioxide if facilitated in the venous blood. About 20% carbondioxide is carried to lungs in this way. At respiratory surface the high partial pressure of oxygen breaks this association and carbon-dioxide is released out.

Carboxy-Haemoglobin. Carbonmonoxide reacts with haemoglobin to form this compound. Affinity of haemoglobin for oxygen is much less than carbonmonoxide, and this substance displaces oxygen on haemoglobin reducing oxygen carrying capacity of blood producing tissue anoxia.

Carcinogen. Any agent that is able to induce transformation in cell to cancerous state.

Carcinologist. Person possessing sound knowledge of carcinology

Carcinology. Branch of zoology that deals with study of crustaceans.

Cardia. Orifice where esophagus enters stomach. *Cardiac*: pertaining to heart. *Arrest*: Due to abnormality in heart beating ventricular extrastoles occur and may lead to ventricular fibrillation due to ventricular tachycardia. The fibrillating ventricles cannot pump blood effectively and circulation of blood stops. leading to cardiac arrest. In absence of emergency treatment. death may occur. within a few minutes.

Carnasial Teeth. In some carnivores (dog etc.). L'le last plemolar in upper Jaw is specially large and used for sheering the flesh from borte; these are termed carnasial teeth.

Carnivorous. Anlmal feeding upon flesh of other animals. *Carnozuria*: group of dinosaurs (Saurischia: Reptilia) that were huge and bulky. about 15 metres long with wide gaping Jaws and sharp stabbing teeth; preyed upon sauropod dinosaurs -*Tyrannosaurus* species.

Carpals. Small skeletal elements of hand arranged in three rows -proximal row bas three carpels (radiale. unlare and intermedium.

middle row has two carpals (centrales) and distal row has five carpals (distalia/distal carpals).

Carpenter Ants. Colonies of ants (Formicoidea: Hymenotera) distinguished by their nesting habits - workers excavate cavities in big plants or tunnel into logs etc. *Bees* : solitary. short-tongued bees *Xylocopa* species (Apoidea: Hymenoptera) characterised by smooth upper surface of abdomen; tunnel and nest into wood.

Carpet Snaks. Small sharks (Pleurotremata: Pisces) characterised by flattened body. broad head and rough mottled skin; found in Pacific ocean. *Snake*. Larg python. *Morella vartegatla* (Ophidia: Reptlla) having variegated skin; found in Australia.

Carpophagous. Fruit-eating

Carpus. Wrtst bones : 8et of bones connecting forearm or foreleg to metacarpus - eight in human being.

Carps. Edible herbivorous fishes which do not have teeth covered with cycloid scales and they have an air bladder Labeo *rohita. Ctrrhtna mrlgala,* Labeo *calbasu* and *Catla* are major carps oi India.

Carrier. Chemical substance which binds together or neutralises a substance to facilitate movement through a membrane. In genetics, an individual whose chromosomes include the trait in question; usually refers to receive trait. Symbiotically, an individual infected with a given. parasite.

Cassowary. Large flightless bird, Casuartus species (Casuariiformes: Neornithes) of northern Australia and New Guinea, about 1.6 metres tall, head black with bony crest; head and neck without feathers (naked) but brightly coloured; wings rudimentary (non-functional), good-runner, feeds mainly on fruits and insects and lays green eggs.

Caste. Any single type of species of social insects (honeybees, termites etc.) that carry our specific function (workers, soldiers etc.).

Castration. Removal of gonads specially testes from an individual. CASURIIFORMES : order of large flightless ostrich-like birds of Australia and New Guinea having no head and neck feathers -Cassowaries (Casuariidre).

Cat. Common name for various feline mammals (Felidre: Carnivora) with digitigrade gait and powerful claws - *Felts domesttca* (domestic cat) predaceous on mice is often kept as pet; others include F. *chaus* (wild cat), *F.caracal* (black cat), F .. *ltbyca (desert cat).F. temmincki* (golden cat) F. *rubigtnosa* (spotted cat) and F. *viverrina* (fishing cat).

Catadromous. Fishes that live in fresh-water and go to the sea for spawning - eels. Opposite to andromous.

Catalysis. Acceleration of a chemical reaction by a substance (called catalyst) that itself undergoes no permanent chemical change.

Catamount. European wild cat: puma *(Felis* species) *Catecholamines*: collective narpe for epinephrine and norepinephrine, found in certain groups of neurons. These are hormones secreted from medulla of adrenal gland; influence various activities by exercising control on sympathetic and parasympathetic nervous systems; also contribute to 'flight or flight· response during emergencies.

Catenulida. Order of colourless, thread-like fresh-water rhabadocoeles (Turbellaria: Platihelminthes) having simple pharynx and single median protonephridum.

Caterpillar. Larva of lepidopterous insects (Hexapoda) with well developed head. three thoracic and ten abdominal segments, pair of five-jointed legs on each thoracic segment and generally five pairs of prolegs; some have, 2, 4 or 8 pairs of prolegs and a few are apodous.

Cat-Fishes. General name for bony fishes (Silurdae) having barbels which look like whiskers of cat. Body naked and fins usually with spines; chain of bones connecting ears with air bladder: some can stay outside water for sometime and can walk for about half a kilometer with the help of their fins; mostly predators - *Clarius batrachus, Amphipnous cuchta* etc.

Cattalo. Cross between domestic cow and bison.

Cattle. Pasture-animals especially cows, bulls, buffaloes and sheep, sometimes. even horses.

Cauda Equina. Bundle of nerves foroled by last four pairs of spinal nerves which become elongated: lies inside the vertebral column along the filum terminale.

Caviare. Salted roe of sturgeon or other large fresn-water fishes. CAVY: any large rodent of *Cauia* species (Cavijdae: Mammalia) including capybara. guinea pig. mara etc.

Cell. Small mass of protopalsm surrounded by a cell membrane and containing one or more nuclei; structural and functional unit or organ!sattonal unit of organisation of living things.

Cellular. Structure consisting of cells.

Central Sulcus. Deep, almost vertical furrow on the cerebrum. dividing the-frontal and parietal lobes.

Centriole. Cytoplasmic organelle in the animal cells which help in formation of spindle during mitosis/meiosis.

Centrolecithal. Egg with yolk confined to centre leading to superficial cleavage. Occurs ini nsects.

Centraomere. Point where the spindle fibres are attached; constricted region of chromosome not coloured by usual chromosomal stains.

Centrum. Ventral portion of vertebra formed by ossification of embryonic notochord form which spring *oui* neural spines, transverse processes· and zygapophyses.

Cephalisation. Tendency. of centralisation of important parts like sense organs and mouth in the head region.

Cephalopoda. Class of highly developed molluscs having well developed head that combines the foot as well; mouth located in centre of foot and surrounded by circlets of arms which are fully equipped with suckers that help in grasping and capturing the prey (usually small fish) - cattlefish, nautilus, octopus etc.

Cere. Patch of a swollen skin at the base of beaks in some birds like Emu.

Cerebral. Brain as a whole or anterior dorsal hemisphere.

Cerebrosides. Compounds containing galactose, fatty acids and sphingosine. Principal constltutent of nervous system.

Cerebrum. Large lobed anterior part of brain which in man makes possible hearing speech memory etc .• seat of sensory power.

Cerulin. Intresting decapeptide isolated from skin of Australian frog. *Hyla caerulta* has some 5 C-terminal and 2 N-terminal amino-acid residues. This substance has all the properties of gastrin and cholecystokinin in mammals and is being used in human beings to contract gall bladder during cholecystography.

Ceruloplasmin. Copper containing plasma protein. Deficiency causes Wilson's disease. a familiar disorder of copper metabolism.

Cervical Vertebrae. Certebae found in the neck region of a vertebrate, like reptiles. birds and mammals. The number is generally 9 in reptiles. 14 in birds and 7 in mammals.

Cervix. Neck-like structure between utelus and vagina in a typical mammal; opening of uterus in a mammal.

Cetacea. Order of large bulky aquatic mammals (Eutheria: hordata) streamlined fish-like body; forelimbs modified into paddles and hind limbs absent, posterior end of body expanded into a flat tail; skin a thick layer of fat (blubber) - two suborders, Mysticeti (toothless-whales) and Odontocetl (dolphins and toothed-whales).

Chaetetidae. Family of Paleozoic corals (Tabulata) Chatodontid: family of fishes having slender teeth (Perciformes: Percoidel) -butterfly-fishes.

Chactognatha. Phylum of abundant planktonic arrow worms. CHAETOPOD : any worm that crawls with help of bristles or setae -earthworm, sea-mouse etc.

Chaetotaxy. Arrangement of principal bristles or setae on insec exoskeleton

Character. Distinguishing structure of function, any trait of an organism.

Chelonia (Testudinata). Order of reptiles (Chordata) that includes many extinct and some living. tortoises (terrestrial), turtles (semiaquatlc and marine) and terrapins (fresh-water) characterised by presence of dermal and epidermal armature. absence of teeth and vertical anus.

Chelosis. Inflamation of comers of mouth; associated with deficiency ofVitamln B_2.

Chemosterilant. Chemical used for making insects sterile without killing them.

Chemotropism. Behaviour response of an organism to chemical stimulation.

Chick. Young fowl, especially hen; young domestic bird

Chitin. Complex organic substance occurring in the exoskeleton of arthropods and some annelids.

Chitons. Group of primitive worm-like mollus having eight overlapping transverse plates covering their backs, feed on algae coating rocky shores and are often seen clinging to the rocks when tide is out.

Chroride Shift. Movement of chloride ions across the RBCs membrane to compensate the movement of bicabonate ions.

Chilocruorin. Green respiratory pigment containing iron found in certain marine polychaetes; similar to haemoglobin.

Chilorogogues. Chlorogogencelis: excretory cells of certain invertebratt's including annelids canylng wastes from body fluid to epithelium .

Chilorophyll. Green pigment in plants and possibly some lower animals that is essential for photosynthesis.

Chiloroplast. Chlorophyll containing romatophore.

Choanocytes. Fiagelleted collar cells found in sponges. Responsible for setting water current through their body.

Cholecystokinin. Harmone liberated by upper intestinal mucosa that induces gall bladder contribution and the release of pancreatic juice. Aleo called pancreozymin (CCK-PZ).

Cholinergic Discharge. Discharge/release of acetylcholine or substances with similar actions.

Cholinergic Neuron. Nerve fibre that releases acetylcholine from axon terminal when impulses are transmitted across the synapses during stransmisston of nerve impulse.

Chondrichthyes. Class of exclusively marine fishes that includes sharks and rays, possessing endoskeleton of cartilages.

Chcrae Tendiane. String-llke cords attached to myocardium wall and atrtoventricular valves. which on concentration of ventricles help to keep those valves shut thereby keeping blood from flowing back into the aurtcles.

Chordate. Any species belonging to phylum Chordata (Animalia). *Chorion*: extraemblyonic membrane found in reptlles, birds and mammals. In mammals it contrtbutes to the placenta. Also outer shell of insects' eggs, used in nutrition.

Choroid Rete. Counter current arrangement of aterioles and venules behihd retina in eyes of bony-fish (generally a teleost).

Chromatid. Either of the two identical halves in which a chromosome appears to split longitudinally in the cell division; actually each chromosome duplicates another copy next to itself.

Chromatin. Stainable protoplasmic substance present in the cell which gives rise to conspicuous chromosomes during mitosis /meiosis.

Chromatogram. Preparation made on papel" with the help of chromatograph for detection of amino-acids present in any sample.

Chromomere. Individual chromatin granules in a chromosome *Chromonemata*: number of thread-like structures (at least four) in chromosome, seen during prophase stage of meiosis. Difficult to see except in forms like protozoa.

Chromophore. Specialised pigment bearing body or cell such as chromatophroe responsible for colour making in many animals.

Chromoprotein. Protein conjugated with pigment *e.g.*, haemoglobin, myqglobin etc.

Chromosomes. Deeply staining structures of nucleus of cell visible under microscope only. Consists of essentially genes arranged in linear order.

Chrysalis. Pupae of butterfly Oepidoptera: Hexapoda).

Chub. Small thick, fat, coarse-fleshed fresh-water fish *Leuciscus cephalus* (Cyprinid Pisces); dusky-green dorsally and silvery-white beneath; found in Europe and Caspian sea.

Chyle. Fluid food within the intestine (the food during passage to intestine receives a large supply of water' brought through various juices rendering it more fluidly which is also due to digestion of various food stuffs).

Chyme. Food in stomach is reduced to thick creamy paste owing to addition of water, churning of food by stomach walls and by action of gastric juices and spurted into duoenum in form of small jets.

Chymotryspin. Enzyme in the pancreatic juice that supplements the renin action.

Ciliary Muscles. Modification 01 choroid coat to which zonules of Zinn attach; secrets aqueous and vitreous humors in the eye.

Cimicid. Family of blood sucking bugs, flat broadly oval body. reddish-brown in colour and without wings; nocturnal and ectoparasites - bedbug etc.

Circle of Willis. In man the internal carotid and vertebral arteries (forming basilar artery) are united at the base of brain at the circle of willis from this anastomosis six cerebral branches go to the cortex, subcortex and brain stem.

Cirrhosis. Diffuse hepatic fibrosis generally caused by fatty liver or either infections. The I chronic condition of fatty liver leads to death of many liver cells and their replacement by scar tissues. The scarring. by interferring with the hepatic blood supply. aggravates the condition leading to more necrosis and scarring. The disease is either severe malnutrition or due to alchoholism.

Cirrus. Small slender projection or appendage appearing almost like tentacle except for its position.

Cisternamagna. Cavity dorsal to coelom of vertebrates in which kidneys are located.

Cistron. Genetic unit of biocherrical function: the sequence of nucleotide pairs in DNA that specify the sequence of a single peptide chain.

Citracacid Cycle. Metabolic cycle responsible for complete oxidation of acetyl portion of acetyl Coenzyme-A. Regenerates 36 molecules of ATP. Also called Kreb's cycle/TCA cycle.

Cladistia. Old name for Polypteriformes. An order of antinopterygian fishes.

Cladocera. Suborder of Branchiopoda (Crustaca), carapace encloses trunk limbs. swims by means of antenre - water fleas *(Daphnia* species).

Cladocopa. Suborder of Myodocopida (Ostracoda: Crustacea) marine forms having carapace without permanent aperture when the two valves are closed - *Conchoecia* species.

Cladocopina. Old and equivalent term for Cladocopa.

Clam. Common name for bivalved molluscs *e.g., Unio. Lamellidens* etc., usually lie partly burled in muddy/sandy bottoms of lakes or streams. Burrow and move from place to place with help of foot which can be extended from anterior end of shell. Water with food and oxygen enters the body through a slit-like opening at the posterior end of shell (incurrent Siphon) and water passes out with wastes and faecal matter etc. through same end of shell (excurrent siphon): to clog or smear: sound produced by ringing together two or more bells.

Claspers. Pair of hard elongated structures attached to pelvic fins of some cartilagenous fishes like *Scoliodon.* These have a groove internally and are connected with cloaca and are used to transfer sperms into cloaca of female.

Climatic Migration. Locomotory activity bringing the organism to ' preferred climatic zone.

Climax Comunity. Ecological community having reached a stat" of dynamic equilibrium.

Climbing Frog. Tree-frog, an arboreal amphibian (Hylidre: Anura).

Cline. Continuous series of differences in structure or function exhibited by members of a species along a line extending from one part of their range to another.

Clitellum. Thickened glandular portion of body of earthworm or other annelid used for formation of coccoon.

Clitoris. Homologue of penis present in female mammals as a small rod-like structure at the entrance of vestibule.

Cloaca. Common passage/way/cavity at the posterior end of body into which intestine, kidneys and genital organs discharge their products; found in Amphibia, Reptiles and egg-laying mammals.

Clypeus. Rectangular sclerite ventral to frons of an insect's head. *Cnidaria*: a group of animals ail possessing cnidoblasts e.g., *Hydra. Cnidoblast*: type of cell the enidarians possess; forms nematocysts.

Coati. Carnivorous, plantigrade mammal, *Nasua nortca* found in forests of North America, 1.4 to 1.6 metres, long having long flexible tnmk-like snout and tail usually held vertical.

Cobra. Group of venomous snake, *naJa* spp. (ophidia:reptilia), about 2 metres or more in length; when excited dilates its neck forming hood-like shape; common in Indian subcontinent and Mrtca, feeds mainly on rodents. *King*: largest species of cobra, *Naja bungarus* that grows upto five metres in length and is deadly poisonous; found in South Asia (including India).

Coccidia. Subclass of telosporea (sporozoa: protozoa). that includes intracellular parasites of epithelial tissue in various animals.

Coccoons. Silken protective cases about a mass of eggs, larvae, pupae or adult animals.

Coccosteomorpid. Aberrant lineage of joint-necked fishes. *Coccyx* : caudal vertebrae of man (3-4) fused into a single mass; a vestlgeal organ in man.

Cochlea. Structure within internal ear of mammals which lodges organ of corti.

Cock. Male of domestic fowl. *Gallus domesttcus* (Galliformes: Aves). as also various other birds.

Cockatoo. Brightly coloured parrots, *Kakatoe* spp. (Psittaciformes: Aves) with movable crest or tuft of feathers on head; common in Australasia.

Cockchafer. Beetle-grub or May-beetle: large stout greyish-brown beetlem *Anomala* spp., *Holotrichia* spp. (Melolonthidae: Coleoptera) both grub and adult live in soil and feed on vegetation.

Cockle. Bivalve *edible mollusc, Cockleedule* having ribbed shell, found mostly on sandy coasts.

Cockroach. Oldest IMng insect (Blattidre: Dictyoptera) that has thrived from Carboniferous period till today (300 mUlion years); earlier ones were 13 to 15 cm long but now these are less than 5 cm in length; brown to black in colour. body broad and flattened. head bent down and concealea from above by pronotum. cerci well developed. noctumal in habit;· common household pest -*Blatta orientalis, Pertplanata* amerlcana etc.

Codon. Nucleotide triplet that ca;t give code for a specific amino acid or acts as termination signal.

Coelenterata (*Cnidaria*). Phylum of Metazoa; two-layered. many-celled animals. having sac-like body cavity and tentacles with stinging cells; - it is assumed that coelenterates originated from ancient colonal Protozoa -

Coenztime. Organic molecule that combines with an apoenzyme to form functional holoenzyme.

Coenztime-a (CoA). Derivative of panthothenic acid to which acetic acid is attached to form acetyl coenzyme - A.

Co-factor. Ari atom. ion, or molecule that combines with an enzyme to activate it.

Cofficient of Consanguinity. Value which expresses the chance that an individual has received from both parents a specific gene which has transmitted to them by an ancestor they have in common.

Cold Blooded. Poikilithermal: animal without ability to regulate body tetpperature by physiological means - fishes, Amphibia and Reptiles.

Coleoptera. Largest order of insects (Hexapoda); biting and chewing mouthparts large, mobile prothorax, foreWings modified into horny or leathery elytra: hind wings membranous; metamorphosis complete -beetles and weevils.

Coliformes. Order of birds (Aves: Chordata) characterised by short legs with four long toes in each, directed forward and a long tail; includes single family Collidae - coolies , mousebirds etc.

Colinearity. Corresponding relationship which exists between order of amino-acids along a polypeptide chain and order of codon in DNA of the gene which specifies that polypeptide.

Collagen. Substance making up parts of tendons; consists of fibres of white variety.

Collecting Duct. Portion of renal tubule in which final concentration of urine occurs; distal part of convoluted part of renal tubule that opens into ureter.

Collembola. Order of small aperygote insects (Hexapoda); ectognathus mouthparts and 6-segmented abdomen with springing organ that enables the insect to Jump into air; metamorphosis slight (2000 species) - springtails and snowflies.

Colloid. State of matter in which particles larger than single molecules are distributed through out a medium such as liquid, gas or solid and which is called dispersion medium.

Colloidal Solution. Solution of fine colloids suspended in a liquid. ***Colon***: .antenor part of large intestine of vertebrate.

Colony. Group of individuals, unicellular/muiticelllar, of same species who have developed from a common parent and remain organically attached together; also used for insect societies. Opposite to solitary.

Colorado Beetle. Potato bug, *leptinotarsa decemHneata* (chrysomelidare coleoptera) ; adult 12-15 mm long, round body, clay-yellow with ten longitudinal black stripes on elytra; major pest of potato in USA since 1823.

Colostrum. Small quantity of first milk after parturition in mammals. (see mammary glands).

Colt Young Horse *(MAIE):* also a young ass or camel

Colugo. Flying lemur (Dermoptera: Mammalia).

Columbiformes. Order of grain eating birds (Carinate: Aves) that are excellent and stylish fliers; young ones are nourished by pigeon's milk secreted in crop glands - doves and pigeons.

Collumellaauris. Rod-like structure extending across the cavity of middle-ear from ear drum to inner ear; serves to transmit the vibrations to inner ear.

Coma. State of prolonged unconsciousness with severe disruption of conscious/voluntary control mechanisms of the body; may be reversible or irreversible; often state of unconsciousness is broken by periods (temporary) of partial consciousness. Cause unknown but generally found in case of severe hyperglycemia/diabetes (Diabetes mellitus) during which blood· sugar rises from 0.1 % to 0.2-0.4% and there is loss of sugar with urine. As a result fat and proteins are subjected to tremendous loss and there is seen weakening along with acidosis of body due to accumulation of ketone bodies in blood (clinically called diabetic coma). Death generally occurs in 2-3 weeks.

Commensalism. Association of two individuals in which at least one is benefited and other is neither benefited nor hanned *e.g.,* sparrow in the house.

Commissur. Nerve connection joining two similar ganglia. COMMUNITY: complex group of plants/animals that occupy a particular area.

Compound Eye. Eyes of insects and arthropods. consist of several ommatidia each of which functions like an independent eye .

Conchostraca. Order of bronchiopoda (Entomostraca: Crustacea) that includes small mussel, like animals -*Lymnetis* species, *Lynces* species.

Conditioned Reflex. Reflex actions which are developed as a result of learning/training *e.g.,* cycling. dancing, swinuning, singing etc. In early phase these are under control of cerebrum but later become independent of it.

Condor. Large South American vulture, *Vultur gryphus* (Falconiformes:aves) blackish plumage and fleshy excrescene falling over its back, wing span three metres or even more; inhabitant of high mountainous regions.

Condyle. Round ball-like structure formed at the extermity of some bones *e.g..* femur, humerus; also occipetal condyle when associated with skull.

Congenital. Present at birth may be due to genetic or solely due to environmental causes.

Conger. Large edible eel that grows upto two metres in length; inhabits sea waters. common on European coasts.

Conjugation. Temporary union of two sexes of same animal for exchange of genetic material or sex cells.

Conspecific. Conspecific term applied to individuals or population of same species.

Constipation. Slowing of rate of movement of faecal matter in the intestine/colon leading to accumulation of hard and dry stool; caused by untimely and improper habits of defecation.

Consumers. In an ecosystem those which obtain energy from producers (primary consumers) or from other consumers (secondary consumers).

Continental Island. Island which once was attached to nearby continent. Normally fauna are characteristiC of that continent.

Continuous Variation. That phenotypic variation seen in many characteristics which does not fall into distinct sharply defined categories but which grades from one extreme to the other.

Contour Feather. Feather with stiff shaft and vane. Appears on exterior of body and gives a definite she to the body of bird.

Contractile Vacuole. Osmoreglatory apparatus that can secrete water from cytoplasm?

Conus Arteriosus. Expanded cone-like structure over the ventricle in some vertebrates *e.g.,* fishes that empties blood into aorta. It has valves that check back-flow of blood.

Copulation. Temporary union between male and female for;-eceiving sperms leading to fert111sation: Coition or intercourse.

Coral Feefs. One of the formation of coral polyps confined to warm waters (South Pacific and Indian· Oceans); Built by epidermal cells of countless small polyps each one secreting in cup-shaped skeleton. Old polyps die and new generations secrete cups upon old ones only; surface of coral mass is alive. There are three types· of coral reefs: ***Atoll reef.*** A belt of coral reef surrounding a central lagoon. These are there in mid Pacific with lagoon of few hundred meters to several kilometers in diameter. Bikni atoll has a lagoon area of 7.25 sq kilometers and land area of 7.43 sq kilometer. ***Barrier Reef.*** It is separated from shore by a wide deeply formed channel (usually affording passage for relatively larger boats) like great barrier reef of Australia (2160 km. in length and channel is 20 to 50 meters deep). It may entirely surround an island. ***Fringine Reef.***: ridge of coral built up from sea-bottom, located so near to land that no navigable channel exists between it and the shore. Frequently breaks occur and irregular pools and channels are created in the reef which are often inhibited by many kinds of animals (often brilliantly coloured).

Cormidium. Assemblage of individuals or organisms found dangling in clusters from main stem of pelagic siphonophores.

Cormoranis. Group of large voracious sea-birds with glossy-black plumage, long neck, hooked bill, strong wings and webbed feet; feed on fishes-*Phalcrocorax carbo* (Pelecaniformes: Aves) tamed and used for catching fish in eastern China and Japan.

Corena. Outer very transparent layer of eye secreted by corneagen cells.

Cornet-fish. Marine fish (Pisces) having slim body and enormously elongated tube-like head.

Cornu Horn. Horn-like part of process ***Corunated*** : having horn or horn-like process ***Corona***: ring/circle of tentacles or cilia.

Coronary Thrombosis. Occlusion in the branch of coronary artery of heart by a thrombus due to plaque formation as a result of arteriosolerosis leading to no-supply of blood to myocardium and may bring about myocardialnfraction. If major vessel is blocked, there may be instant death.

Corpora Cardiaca. Neurohaemal organ in insects situated between brain and corpus allatum. Neutrosecretory axons extend from brain to this part. It secrets its own hormones but serves mainly to store and release reurohormones from brain. Its hormones accelerate heart beating and peristalsis in the fore gut. They also regulate protein metabolism in fat body and stimulate secretion of moulting hormones.

Corpora Cavernosa. Two lateral columns of erectile tissue in penis of mammals separated by a septum in the middle attached to ischium or pubis. A small penis-bone or ospenis is present in septum of these columns:

Cor Pulmonalf. Due to degenerative pulmonary disease alveoli of lung are replaced by large air sacs as a result of inspiration and expiration are laboured and breathing is greatly increased. Pulmonary hypertension develops and right side of heart enlarges and then fails; the enlarged side is called cor pulmonale.

Corpus Albicans. If pregnancy fails, the corpus luteum regresses and is replaced by scar tissue called corpus albicans.

Allatum. Endocrine gland behind brain in insect's head which secretes hormone associated with larval moults.

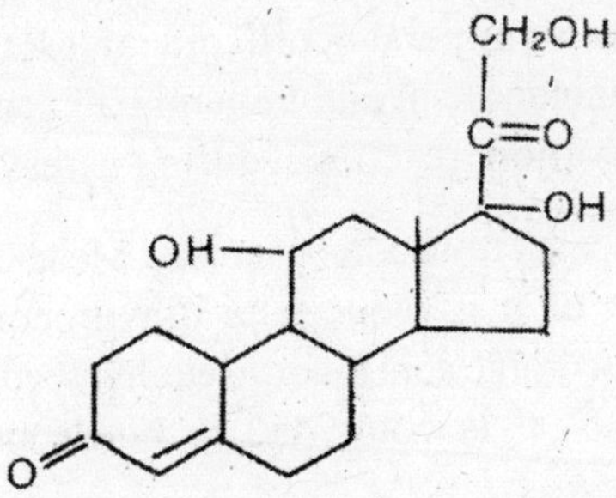

Corpuscles. Cellular structures of blood, constituting 40% of it. Three types can be disintinguished - RBCs, WRCs and platelets (RBCs 400000-500000/ml. WBCs 7000/cu mm and platelets 250000/cummm).

Corpuscular Badiation. Stream of subatomic particles which can travel through space and dissipate their energy upon contact with matter. '. .

Corpus Haemorrhagium. After release of ovum from Graafian follicle the entire follicle with little blood inside it, may develop to corpus luteum, if pregnancy has occured. but if pregnancy fails it develops into scar-tissue called corpus haemorrhagicum. ***Luteum*:** After release of mature ovum whatever is left in Graafian follicle: endocrine organ secreting progesterone in mammalian ovary; develops as corpus luteum of pregnancy if ovum gets fertilised and secrees hormones that are responsible for maintenance of pregnancy; if pregnancy fails. then regresses: activity is controlled by luteotropic hormone (LH) from anterior pituitary. ***Spongiosum*:** One of the three columns of erectile tissue in penis of mammals, immediately surrounding urethra. Enlarged at anterior end to form most sensitive part called glans. ***Stratum*:** Thickened ventrolateral sides of floor of cerebral hemispheres.

Cortex. Outer portion of a structure of an organ of gland - adrenal cortex.

Corticosteroids. Steroids derived from cholesteral secreted by cortex of adrenal gland; include glucocorticoids like corticosterone, II-Odeoxycorticosterone andmineral ocorticoid .aldosterone. Control carbonhydrate metabolism and mineral balance also associated with conservation reactions during stress.

Cortisol. One of the glucocorticoids from adrenal cortex. Metabolised in the liver where most of it is reduced to dihydrocortisol which is conjugated to glucuronic acid to form tetrahydrocortisol glucuronide and some of it is converted to cortisone.

Cortisone. Drug which may stabilise membranes: formed in the body by conversion of cortisol within the liver.

Cotingas. Group of tropical passerine birds. *Cottnga* spp. (Cotingidae: Aves) With brilliantly coloured plumage, live mainly in rain forests of Centra) and South America.

Coughing. Body disturbce caused by irritation of trachea or large bronchii. The epiglottis in this process closes glottis tightly and air is forced through tubes which results in clearing off the blockade within the tubes. It is a reflex response to irritation of receptors in mucosa of large respiratory passages.

Cow. Female- of any bovine animal especially domestic species. Bos tndtcus (Artiodactyla: Mammalia). female of elephant. rhinoceros. seal. whale etc.

Cowper's Glands. Paired glands producing mucous which help in passage of sperms down the urethra: also called bulbourethral glands. Present in male and female mammals only.

Coyote. Prairie-wolf *Canis latrans* (Carnivora: Mammalia) small wolf-like animal found in western North America; preys upon rodents.

Coypu. Large aquatic rodent. *Myocastor coypus* (Rodentia: Mammalia). found in South America; valued for its fur (nutria).

Crab. Aquatic arthropod. Carctnus spp. and allied forms (Decapoda: Crustacea); broad. flat exoskeleton or carapace; eyes stalked and movable; abdomen short and bent under throax; five pairs of legs, first pair being larger ends in nippers or claws; always moves sideways; mostly marine. some live in fresh-waters as well.

Cracid. Family of game-birds (Galliformes: Aves) common in New World tropical upland - curassows, guans etc.

Cranial Nerves. Nerves consisting of sehsory,motor or both fibres coming out of brain. Their number being 10 pairs in fishes and amphibians and 12 pairs in mammals. Their names are -olfactory, optic, occulotnotor, pathetic, trigeminal, abducens, facialis, auditory, glossopharyngeal and vagus. In addition spinal accessory and hypoglossal are found in mammals only.

Cray-fish. Common name for crustacean like *Astacus* species -prawn, cambarns etc; fresh water forms and edible-also spiny lobs tar, *Palinurus vulgaris.*

Craw. Crop or first stomach of birds and insects

Creatine Phosphate. One of the muscle phosphagens which . liberates energy-rich $P0_4$ - hond to provide energy for muscle contraction.

Cribriform Plate. Part of ethmoid bone bearing numerous small foramina for olfactory nerve.

Cremaster. Hooked or spine-like appendage at anal end of lepidopterous pupae.

Creodont. Primitive Eocene fossil carnivore mammal showing affinity with members of Carnivora and Insectivora orders; having large canine teeth but no camassials.

Crepuscular. Animals active in twilight, pre-dawn and dusk. *Cribellum*: accessory spinning-organ of certain spiders (Arachnida) and some insects (Hexapoda).

Cricoid. Ring-like cartilage supporting larynx in vertebrates. *Crickets*: saltatory orthopterous jumping and leaping insects, *Acheta* spp. (Gryllidre: Tettigonioidea); produce characteristic shrill noise by rubbing together parts of forewings —*A domestica* is common house cricket.

Crinoidea. Class of radially symmetrical crinozoans; adults flower-shaped with anchord stem or free-living - *Antedon* species. *Hybocystites* spacies etc.

Crinozoa. Subphylum of Echinodermata; radially symmetrical creatures that show partly meridional pattern of growth -Crinoidea and Heterostelea.

Crista Ampullaris. Receptor structure located in the expanded end of ampulla of each of membranous semicircular canals of ear.

Cristae. Membraneous formation in the inner wall of mitochondrion which is raised into projections; projections formed by-folding of inner mitochondrial membrane into its matrix. Thickness varies from 180 to 200 A.

Crocodiles. Group of large scaly reptiles (Crocodylidare) inhabiting rivers and estuaries in warmer parts of World, but lays shelled eggs outside water; upper part of body dark olive-brown with spots or bands. snout long. and pointed, teeth unequal in size; ferocious, feeds on fish. birds and small mammals that come for a drink or swim; longevity about 100 years - *Cricodilus palustris* and C. *porosus* are Indian species. C. acutus (U.S.A.) and C. *nilottcus* (Africa).

Bird. Small Egyption black-headed plover *Pulvianus aegyptius* (Charadriidae: Aves) ; having lavender and cream plumage with black and white markings; feeds on crocodile's insect parasites.

Crocodilia. Order of largest lying reptiles (Chordata) that are amphibious; body covered wltp rows of sculptural bony plates and epidermal scales; fore-and hind limbs adapted for walking while .tail helps in swimming - alligators, caimons, crocodiles and Gharials *(gravial* species).

Crochets. Hooked spines on tip of prolegs of caterpillars; used for crawling.

Crop. Bag/sac like structure formed in certain animals like insects and birds; serves to store ingested food material.

Crop-Glands. Pair of glandular structures present in the crop of birds; during breeding season they secrete 'pigeon's milk' in both sexes under influence of prolactin hormone from pituitary.

Crossbil. Parrot-like finch. *Loxia* .species (Psittaciformes: Aves). characterised by crossed tips of beak; male bright red with black spots, female olive green found in coniferous forests of northern homisphere.

Crow. Common name for moderately large passerine black birds(s), *Corvus* spp. (Carvidae : Aves) - common Indian species is C. *varius* while C. *corone* is English carrion-crow.

Crus Commune. Common duct formed by proximal parts of anterior and posterior canals of internal ear in mammals.

Cryptorchism. Condition when testes in a baby do not descend in scrotum and remain undescended. Defect may be treated with gonadrotropic hormones or surgery.

Cuculifrmes. Order of perching birds (Carinatae : Aves); distinguished by having zygodactlyl toes. characteristic sound notes and habit of laying eggs in nests of other birds - cuckoo. koels etc.

Cupula. Gelatinous mass on tips of sensoty hairs of lateral line sense organ in fishes.

Curassow. Groups of turkey-like birds (Aves) characterised by horny helmets. formed in Central and South America; feeds on fruits and leaves.

Curculionid. Family of curculionoidea (Coleoptera : Hexapoda) that includes all true weevils.

Cursoria. In old classification of insects. refers to running Orthopterans such as cockroaches. mantids and walking-sticks.

Cushing Syndrome. Clinical picture produced by excess of glucocorticoids. Persons suffering are characterised by protein depletion. poor wound healing. poor muscle development. thin skin, redish-purple striae, red cheeks. moon face fat pads (buffalo hump) and pendulous abdomen.

Cutaneous. Thin noncellular outermost covering of an organism. *Cuticula*: more ·modem form oc cuticle preferred because cuticle implies inactivity.

Cuticulin. Substance making up epicutlcle.

Cutis. True skin as distinguished from cuticle (outermost non-cellular layer of body wall).

Cuttlefish. Marine mollusc *Sepia offlcinalis*. distinct head, two well developed eyeseight short and two long tentacles; three hearts and well developed shell; when disturbed, ejects black inky fluid so as to darken the surrounding water as means of protection.

Cyanacoralamine.. Vitamin-B_{12} : water soluble vitamin synthesised by bacterium *streptomyces olivaceous* under effect of $C_0$60 radiations. It is found in kidney, liver and meat. Essential for growth and proper erythropoisis (RECs formation). Deflciency causes pernicious anaemia. Daily requirement 2-3 J1.g.

Cybernetics. Science of information, communication and control in animals and machines.

Cyclic-amp. Compound derived from ATP by ~denylate cyclase that serves as a second messenger for hormonal action; hormone association with membrane receptor leads to synthesis of c-AMP from ATP. c-AMP then removes inhibition of protein kinase cy removing inhibitory sub unit; active protein kinase can phosphorylate another intracellular enzyme thereby either activating or repressing it depending upon enzyme. Since there are various forms of kinase with different substrates that they phosphorylate. the hormone induced increase in c-AMP concentration can lead to any variety of effect.

Cycli GMP. Another regulatory nucleotide, present in cells at a concentration lower than c-AMP and produced from GTP by guanylate cyclase. It promotes responses opposite to those promoted by c-AMP.

Cyclophilid. Old name for order Taenioidea (Platyhelninthes).

Cyclosis. Rotary slow streaming movement of protoplasm - certain cells and animals species k. *Paramecium*

Cyclostmates. Subclass of Agnatha that includes primitive craniates with circular mouth having no jaws; body long, cylindrical; skin smooth; cartilagenous endoskeleton, gills in pouches, 5-16 pairs; heart two chambered-lampreys and hag-fishes.

Cyst. Thickened resistant wall: a sac or bladder-like structure surrounding or enclosing an organism.

Cysticercold. Larval stage of some cestodes like. *Hymenolepis* nana found in body of certain insects - dung-beetle.

Cystoporata. Order of extinct marine bsyozoans distinguished by cystopores and minutopores.

Cytheracea. Superfarnily of ostracods (Podocopa : Ccrustacea) that includes very small crawling and digging creatures.

Cytogenctics. Scientific study that links structure and behaviour of chromosome with inheritance.

D

D- 600. Organic compound (drug) that blocks calcium influ.x through cell membranes. Chemically the substance is alphaisopropyl alpha Vitamin D_3 (N-methyl-N-homovertaryl)-alpha-amino propyl)-3, 4. 5-trimethoxyphenyl acentnitrile.

Dab (Dabster). Small. brown. flatfish. *Limanda limanda* (Pisces) found in shallow coastal waters. about 40 cm long. characterised by sharp bend in lateral line just below gill cover.

Dactylognathite. Distal segment of maxilliped crustaceans. DACTYLOGROIDEA : superfamily of Monogenea (Trematoda: Platyhelminthes), liver flukes that are ectoparasites of fishes.

Dactylogist. Person who can talk or explain everything by figure-signs.

Dactylology. Art of talking or explaining with fingers.

Dactylopiidaje. Family of cochineal scales (Coccoidea: Homoptera) - *Dilctlylopins* spp., -yield cochineal, a dye used for colouring foodstuff and cosmetrcs.

Dactylopores. Small openings on surface of corals (Milleporina; Hydrozoa), through these pores polyps (dactylozoids) protrude.

Dactylopteridaje. Only family of Dactlopteroidei (Perciformes: Pisces) that includes flying gumards. *Trigala* spp. (marine fishes).

Dactylosopidaa. Family of Blennioidei (Perciformes : Pisces)' that includes small tropical and subtropical fishes found mainly in or around coral-reefs - sand stargazers.

Dactylozooid. Sensory polyp found in hydrozoan (Coelnterata) -*Obelia* species.

Daddy-long-legs. Common name for various crane flies (TIpulidce: Diptera) and some spiders (Arachnida) ; all having thin, slender, long legs.

Dameselfish. Group of small brightly coloured fish (Pisces) found in tropical waters especially near coral-reefs. *e.g.*. clownfish that shatters among the stinging tentacles of sea-anemone.

Dark Current. Steady sodium current leaking into a visual receptor cell at oliter segment. The sodium is pumped out actively at inner segment completing the circuit. It is reduced by photoexcitation.

Darter. Snake-bird. *Anhinga melanogaster* (Pelecaniformes : Aves) Small brilliantly coloured fish (Pisces).

Darwinism. Theory put forward by Chales R Darwin in 1858 postulating that species of living organisms originate and evolve through the process of natural selection in their struggle for existence in particular environmental conditions of their habitat. It further postulates that out of a large variety of members in a population only those carrying the traits suitable for maximising the success of survival are chosen through natural selection during evolutionary course.

D-Cells. In human beings 1-8% cells of islets of Langerhans are these cells which contain somstostatin (substance that acts on anterior pituitary and also inhibits seretion of insulin. glycogen and gastrointestinal hormones) and also contain gastrin.

Deamination. Removal of NH2 group from an amino-acid antl leading to the formation of ammonia. Done in the liver of vertebrates during catabolism of amino-acids. *e.g.,* alanine + 1/2 02./E Pyruvic acid + Ammonia. The process is cateysed by specific amino -acid oxidase.

Decacanth Embryo. Ten-hooked embryo found, in tape-worm *Taenia saginata* and infectious to secondary host.

Decapoda. Order of Malacostraea (Crustaca) having carapace well developed, coveting thorax completely - crabs, lobsters, prawns etc.

Decapon. Ten footed animals-crustaceans and arachnids. **DECEREBRATION:** experimental elimination of cerebral activity by section of brain stem by interrupting blood supply to the brain region.

Deciduate Piacenta. In most of mammals *e.g.,* cow, sheep, goat, dog, cat and primates, because of more intimacy between foetal and maternal parts of placenta some of maternal tiss'ue comes out with foetal part at birth. This type of placenta is called deciduate placenta.

Deciduous Teeth. Diphyodont teeth: teeth that fall off dUring early life - milk teeth of mammals, these include incisors. canines and premolars.

Decomposers. In ecosystem those which obtain energy from dead tissues of consumers; usually comprise of bacteria and fungi.

Degenerate Code. Genetic code in which two or more codons may represent the same amino -acidds.

Deglutition. Swallowing of food; initiated by voluntary action of collecting the oral contents on tongue and propelling them back into the pharynx.

Dehydroretinol. (Vitamin-A2). Form ofvitam1n-A occurring in the liver and retina of fresh-water fishes, some invertebrates and amphibians.

Delayed Lethal Gene. Lethal gene affecting the carrier after attainment of reproductive age hence it is the only type of lethal gene that may be transmitted.

Deletion. Chromosome anamoly in which a segment of a chromosom(; has been lost.

Deletion Mutation. Genetic mutation where one or more codons are knocked; shorter protein.

Dementia. Loss/atrophy of brain functioning. Associated generally with senile degeneration in old age. May be temporarily associated with vitamin B deficiency.

Demoiselle. Numidian crane *Anthropoides virago* (Gruiformes: Aves) with long feathers and white plumes behind the eyes: damsel or young graceful lady: a beautiful fish, *Pomacentrus* species (pisces).

Denaturation. Alteration/destruction of normal nature of substance by physical or chemical means.

Dendrites. Fibres of a neuron that conduct impulses towards cyton.

Denervation. Destruction of nerve to any part of body leading to atrophy of the organ. In skeletal muscles it causes muscular atrophy and abnormal excitability of muscle and increases its sensitivity to acetylcholine. Fine irregular contractions of individual fibres (fibrillations) appear.

Dentary. Upper bone of lower Jaw of vertebrate

Denticle. Small tooth or tooth-like projection.

Denticulate. Bearing a row of small teeth or denticles.

Dentine. Calcareous substance accounting for major portion of teeth and placoid scales. Compare with vitrodentine.

Dentitioin. Complement of teeth or general masticatory apparatus.

Deoxycorticosterone. Honnone synthesised and released by adrenal cortex that regulates salt and water balance.

Deplasmolysis. Increase in intracelluar pressure, usually accompanied by cell membrane bursting. Occurs when cell is placed in hypotonic solution.

Depolarisation. Reduction' in or revesal of potential difference that exists across cell membrane during resting stage.

Dermanyssid. Family of order Acarina (Arachinida: Arthropoda); includes mites that are parasitic on poultry and small rodents - chicken mite. Dermanys *us galltnae*

Dermaptera. Order of long slender insects (Hexapoda) characterised by forewrings modified into short leathery tegmina; hind wings mambranous and cerci modified into heavily selerotised forceps. metamorphosis gradual - earwigs.

Dermattis. Inflamatory development/skin troubles generally associated with deficiency of biotin. pathothenic acid (vitamin B complex).

Dermatome. During development of embryo. differentiation of somites, the outer wall of myocoel becomes dermatome which spreads out Just beneath the epidermis and gives rise to dermis.

Dermoptera. Order of nocturnal insectivorous mammals (Eutheria: Chordata) having characteristic fold of skin extendign along either side of body from neck to tal enclosing the limbs- flying lemurs.

Detrimental Gene. Gene that imparts a disadvantageous effect to individual that possesses it. The carrier is hampered in some way by the gene so that changes of reproduction or survival are less than in those individuals as are free from the gene.

Deuterostomia. Animals in which blastopore forms anus second embryonic opening forms mouth opposite the anus.

Deuterotoky. Parthenogenetic reproduction in which females produced offsprings of both sexes.

Diabetes Insipidus. Physiological abnormality characterised by. passsing of large amounts of dilute urine (polyuria) and drinking of large amounts of water/fluids (polydispia). In sufficient secretion of ADH (antidiuretic hormone) is responsible for less absorption of water from glomerular filtrate which passes out in urine.

Diametes Mellitus. Abnormal condition in human beings characterised by exessive sugar in blood and urine and is caused by insufficient secretion of insluin hormone from islets of Langerhans.

Dialysis. Process by which molecules of different sizes are separated due to their selective diffusion through a membrane or some other substance; clinical application of this process is seen in removing urea from blood of uremic patients by passing their blood through a bath of suitable composition in an artificial kidney.

Diapedesis. Process by which many of the neutrophils enter the tissues insinuating themselves through the walls of capillaries. Both the plasma proteins and leucocytes pass through endothelial system and in most cases injury causes leakage of plasma proteins long before emigration of leucocytes takes place. The escape of plasma proteins is a purely passive affair and is partly true for leucocytes which. however. show active amoeboid movements in this process.

Diaphragm. Sheet-like muscle forming a partition between the thoracic and abdominal cavities of mammals; a dividing membrane.

Diaphysis. Long main body of a bone as differentiated from epiphysis.

Diapophysis. Transverse process arising from neural arch but often from centrum; an articular surface serving for attachment of tuberculum of biheaded ribs.

Diarrhoea. Acceleration of movement of chyle through large intestine. caused generally due to infection or due to worries etc. Severe diarrhoea is debilitating and can be fatal in infants. Large amount of sodium. potassium and water are passed out along with stool leading to dehydration. hypovolemia and eventually shock and cardiovascular failure.

Diarthroses. Movable joints; perfect and movable joints between bones. These may be : *Ball and socket* : enarthroses: one bone has rounded end and other a cavity to receive it; It has greatest degree of movement joints of humerus with glenoid cavity and that of femur with acetabulum.

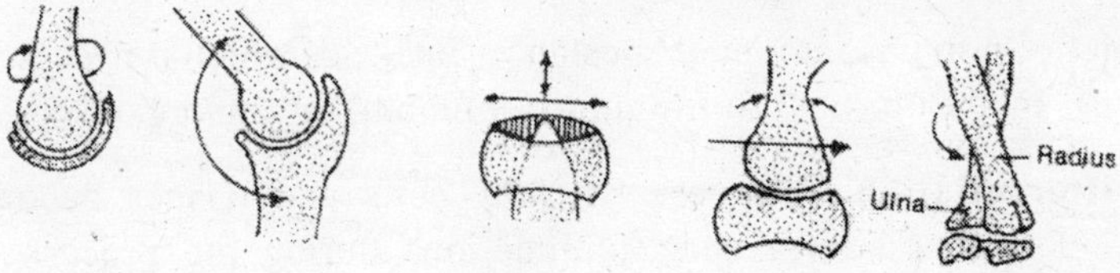

Bail and Socket Hinge Rotary Saddie and sliding

Diarthrosis

Hinge: ginglymus; same as ball and socket but movement is restricted - elbow joint. *Pivotal:* rotaria: one bone is shaped like fixed pivot and other rotates on it - on dontoid process of axis rotates the atlas toghether with head. *Sliding*: arthrodium: in which two articulating bones can slide over one another *Le.,* joints in zygapophyses of vertebrae and in pronation (palm facing downwards) radius is on thumb side and ulna on little finger side. *Saddle* : like ball and socket but since both ball and socket do not develop, the movement of joint is possible in many planes *e.g.* joints between metacarpals of thumb and wrist bones of man

Diastole. Relaxation phase of auricles or ventricles when they fill with blood in protozoa; the formation of contractile vacuole. Opposite to synstole.

Diatryma. Oldest and largest NorthAmerican ground-bird (Diatrymiformes : Aves), Oligocene (Tertiary) fossil, had rudimentary (functionless) wings, huge beak and powerful legs and was voracious flesh-eater.

Diatrymiformes. Avian order (Chordata), includes large flightless birds that had large head and beak, very small wings, and massive legs; all are extinct by now.

Dielectric Constant. Measure of the degree to which a substance is able to store electric charge under an applied voltage: dependent on charge distribution within molecules.

Diencephalon. Region of vertebrate brain Just posterior to cerebrum in adult. In embryo second part of two divisions of prosencephalon (fore brain).

Diet. Optimum amount of substances intake of which is necessary to the normal functioning (optimum) of an organisms.

Differentiation. Process whereby cells and tissues become specialised for specific functions during the process of development and growth.

Diffusion. Movement of molecules from region of higher concentration to lower concentration brought about as a result of their kinetic energy.

Digestion. Chemical conversion of natural food materials (carbohydrates. proteins and fats) to body food materials (monosacharides. amino-acids and glycerol + fatty acids).

Dikinesis. Stage during prophase of meiosis when chromosomes become thickened and chiasmata move to their ends.

Dimetrodon. Permian fossil dinosaur (Pelycosauria: Reptitia) of southern USA, flesh-eating. over 3 metres in length. had dagger-shaped stabbing and seizing teeth; neural spines of vertebrae enormously elongated forming huge sail over the back.

Dimorphism. Difference in size. structure. form. colour etc. between two types of individuals of same species.

Dingo. Wolf-like wild dog of Australia. placental (non-pouched) mammal. about 50 cm high, having erect pointed ears and bushy tail. lives in packs (groups) hunting kangaroos, sheep and wallabies.

Dinitrophenol (DNP). Any of six isomers lnat act a&, aerobic metabolic inhibitors by virtue of their ability to uncouple Oxidation from phosphorylation in mitochondrial electron transport. The chemical formula being C6H3 (OH) $(N0_2h$

Dinoceras. (deinoceras): Large Eocenea poch fossil. stump-footed. ungulate mammals of North America; having three pairs of protuberances on skull

Dinothere. Huge quadruped animal. *Dinotherium* species (Proboscidea: Mammalia) resembling elephant. extinct since long.

Dioctophytida. Order of parasitic nematodes (E'noptia: Nematoda) characterised by peculiar structure of copulatory bursa of male.

Dioecious. Unisexual: having male and female reproductive organs (flowers) in separate plants.

Dipeptidase. Enzyme that can hydroplyse dipeptides to amino-acids.

Diphyodont Dentition. Type of dentition when there are two sets of teeth during life time of individual; found in mammals where incisors, canines, premolars are deciduous teeth and represented in milk dentition.

Diplodocus. Largest and heaviest of all fossil dinosaurs (Saurischia: Reptilia) of North America, herbivorous. grew upto 25 metres in length weighing about 8 tonnes. quadruped having very small head. Long neck. weak jaws with sharp cutting teeth. tail very long and heavy.

Diploid. Number of chromosomes in somatic cells; double the number of chromosomes that occur in a mature sperm/ ovum *(i.e.,*2 x number of chromosomes).

Diplotene. Stage in prophase of first meiotic division when mutaal attraction of bivalents is replaced by repulsion commencing at centromeres. The chromatids at this stage are clearly visible and as the chromosomes separate. it becomes apparent that these are held at various points called chllasmata.

Dipole. Molecule with separate regions of net negative and net positive charge so that one end acts as opposite pole and other as negative pole.

Dipper. Water ouzel: small song bird. *Cindus cinclus* • living by rocky streams; can dive and walk under water.

Diptera. Order comprising true-flies (Hexaboda) having piercing and sucking mouthparts, two transparent membraneous wings attached to mesothorax, hind wings reduced to halteres housefly, fruilly, stemfly etc.

Discoidal Clevage. Incomplete cleavage also called meroblastic cleavage; occurs in eggs having large amount of yolk.

Discontinuous Variation. That phenotypic variation in some characteristics which can be recognised by two or more d1stinct traits or clear cut difference.

Discordant. Term generally used to indicate that two members of a twin pair differ with respect to expression of a given trait.

Disjunction. Normal separation of chromosomes at anaphase of mitosis / meiosis.

Dispermy. Entrance of two & permataozoa into one ovum.

Disruptive Selection. Selection against the distribution, mean for extremes resulting in divtersifing traits.

Dissimilation. Disintegration of protoplasm principally by oxidation.

Dissogeny. Having two sexually mature stages in life-cycle of an individual - larva and adult both sexually mature.

Distal. Away from point of attachment - hand is distal part of arm. Opposite *of* proximal.

Diving Beetle. Fairly large (40 mm. long) beetle, Dyttscus marginaits (Coleoptera: Hexapoda), foudn in ponds and stagnant waters.

Dizygotic Twins. Fraternal twins: twins derived from two separate eggs; each fertilised by a separate sperm: may be of same sex or different sexes.

DNA (*Deoxyribose Nucleic Acid*). Substance of which genes are made and determines inherited characters; consists of two polmuucleotide chains, forming a right-handed helix with one coil positioned more or less underneath. There are four types of nucleotides in polynucleotide chains although each chain may be tousands of nucleotides long. EAch nuclotideconsists of deoxyribose sugar, phosphoric acid and a base which may be purine (adenine, guanine) and pyrimidine (cytosine, thymine). The sequence of their pairing is, ademine with thymine and guanine withcytosine. Teh two polynucleotide chains are antiparallel Bases are on the inner side of helix. By formation of hydrogen bonds two helices are held together. DNA produces copies of itself with great accuracy so that genes (polynucleotide sequences) are passed unaltered from one generation to other.

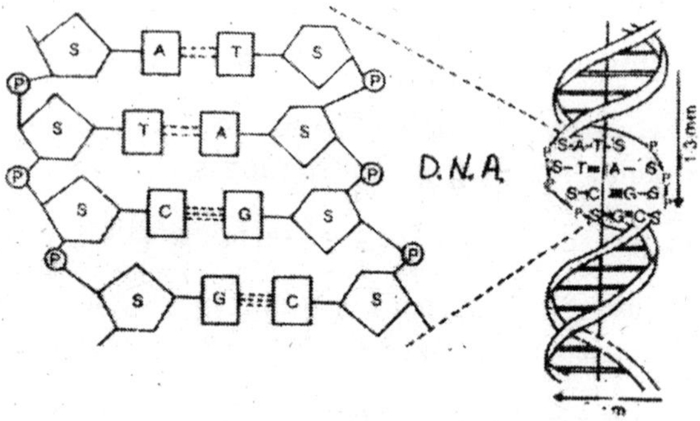

Deoxyribose Nunclic Acid (DNA)

DNA Hybridization. Laboratory technique for measuring the similarity of one of the 1;wo species. DNA from each species is denaturated by heat into single strand and a mixture of DNA strands is incubated to allow them to recombine into hybrid DNA- a process called annealing. The amount of annealing is . directly proportional to similarity of DNA strands and can be measured by heating the hybrid DNA and recording the temperature at which the standards separate.

Dodo. Massive (turbkey-seized). clumsy flightless bird .. *Raphus. cucullatus* (Columbiformes; Aves). native of Maurititus. extincl since end of 17th century.

Doe. Female of buck or fallow-deer. hare or rabbit.

Dog. Carnivorous quadruped. *Canis* spp. (Fissipedia: Mammalia) said to be descendent of wolf *(lupus)* domesticated. semi-domesticated and wild races and breeds. varying greatly in shape. size and colour. are found all over the world.

Dolphin. Sea-mammal *Delphnus delphis.* 2 to 3 metres long. with beak-like snout. shiny-black above and dull-blackish beneath.

Dominant. Species that is abundant or controls major portion of energy flow of a community.

Donkey. Domestic ass.

Dormouse. Old World arboreal rodent. Glits glits (Gliridaae: Rodentia) with bushy tail,it lloks in between rat and squirrel.

Dorsal. Pertaining to back; opposite to ventral in bilaterally symmetrical animal. *Aorta*: main aortic crches (systemic) after passing outwards unite to form a single vessel which runs along the mid dorasl line of body and supplies blood toall visceral organs. *Lamina*: groove-like structure foudn in pharynx of many protochordates like, consisting of 20-30 tongue-like languets that hng doen in the bronchal cavity; covered with cilia, these help to pas food in oesophagus. *Root*: nerve trunk that entrs spinal cord hear its dorsal surface and contains sensory neurons only. *Tubercle*: Chemosensory organ foudninurochordates so consists of a broad base from which arises two spirally coiled cones. In each ocne passes a spirally coiled channel making three oils. Channels of two sides meet at the base. Entire surface is made of ciliated epthelitum supplied by nerve from the ganglion.

Dopa (Dihydroxtphenylalaninej. Intermediate substance in biosynthesis an immediate precursor of nor-epinephrine.

Dove Grain-eating. medium-sized bird. close relative of pigeon (Columbire: Aves) with short neck and legs - brown-dove *Streptopella senegalensis*. ring-dove S. *decaoto*.

Dove's Foot. Various species of crane's-bill Gernium dissectum. G Molle etc.

Downs Sydrome. Mongolism: on an average on child in 700 children has three autosomes on twenty first instead of two chromosomes i.e. 2n+1 (21) = 47 chromosomes. The bearer is charaterised by small body size, retarded/poor mental development, thick tongue; blunt toes, open mouth, oblique eyes and folded eyelids.

Ducks. Common name for various water fowls. *Anas* spp. (Anatidre: Anseriformes) - two main groups; fresh-water surface feeding ducks such as pintail. teal etc .. and sea or diving ducks that include elder, pochard etc.

Duct Tube other than lymph tic/blood vessel through which liquid or other secretion of metabolism from some gland is carried.

Ductless Glands. Endocrine glands which have no ducts to carry out their secretions. A number of such glands *e.g.*, pituitary. thyroid. thymus, adrenals etc. are found in the vertebrates.

Ductus Botali. Connection between 6th aortic arch and portion of dorsal aorta; present generally in embryonic stage. Also called ductus arterosus. *Caroticus*: portion of dorsal between 3rd and 4th arches; present generally in embryonic stage *Cholidochus*: bile duct. *Communicans*: in bony fishes having air bladder e.g., labeo anterior and posterior chambers of air bladder are connected together by this conection and itis controlled by a sphineter. *Cuveri*: in cartilagenous fishes e.g., sharks, the sinus venosus receives blood by a pair of large lateral veins clled ductus cuveri.

Dugong. Seal-like mammal (Sirenia) akin to manate; about 3 metres long. having flipper-like fore-limbs and broad paddle for tail; found in swampy tropical Indian ocean; herbivorous, feeding on el-grass and other aquatic vegetation.

Dugongidae. Family of Sirenian aquatic mammals characterised by elongated sharly deflected premaxillae and no nasal bones; only to species foundin Red Sea and tropical Indian ocean.

Duiker. Small grey antelope of Africa having arched back and short legs; both sexes with long straight horns.

Dung. Beetle, robust thick-set beetles. *Aphodius* spp. (Coleoptera: Hexapoda) having very powerful spiny legs adapted for digging; eggs laid in dung of domestic ani!nals and feed on grubs same dung.

Dunnock. Hedge-sparrow - not sparrow but small bird (Prunellidae: Aves) with long slender beak; found in Asia and Europe; feeds on insects and small grdins.

Duodenum. First part of small intestine posterior to stomach.

Duramater. Outermost membrane covering the brain and spinal cord in vertebrates.

Dwarf. Diminitive person or plant of abnormally low height.

Dwarfism. Abnormally small size in humans; result of insufficient secretion of pituitary hormone during childhood and adolescence. Opposite to giganitlsm.

Dynin Arms. Projection of tubule-A of one microtubule doublet towards tubule-B of next; composed of protein exhibiting ATPase activity.

Dyschezia. Constipation in which faeces arrtve normally at pelvic colon but final evacuation is not adequately performed. If the call to defaecation is neglected, the rectal wall relaxes and desire passes off. The rectum becomes in time excessively distended and contractibility of its wall is impaired and defaecation becomes difficult even when more/great effort is made. Fragmentary defaecation may occur but faeces are left behind in rectum to continue distending it.

Dysfunction. Abnormality or impairment of functioning of an organ of the body.

Dysgammaglobulinemia. Disturbance in the level of gamma-globulins of blood. Caused by abnormality of fat transport in ind1v1duallymphatics.

Dysmenorrhea. Abnormality of menstruation in human female characterised by painful menstruation.

Dysodonta. Order of extinct bivalve molluscs that had toothless hinge and ligament in grooves or pits.

Dysphagia. Difficulty in swallowing.

Dysteleology. Study of functionless, rudimentary organs in animals and plants.

Dzeren. Mangolian antelope, *Antelope* species (Artiodactyla: Mammalia).

E

Eagle. Broad-winged, diurnal, bird of prey, with strong curved beak and powerful claws.

Earthworm. Annelid worm (Oligochaeta), long, narrow, tubular, brown-coloured body, divided into 100 to 120 segments; hermaphrodite; lives in soil.

Earwig. Dermapterous insect (Hexapoda) having pair of forcep-like horny cerci at anal end of abdomen.

Eccrine Glands. Gland in which the products collect at outer part of cells and cytoplasm is broken down before the products are discharged *e.g.,* sweat glands.

Echidna. Spiny-ant-eater. *Tachglossus aculeatus* (Protetheria: Mammalia) found in rocky regions of Australia; primitive, egg-laying, toothless, burrowing nocturnal animal with long extensile tongue.

Echinacea. Suborder of Euechinoidea (Echinoidea: Echinodermata) having keeled teeth. rigid test and branchial slits.

Echiniscoidea. Suborder of Heterotardigrada (Oncopoda: Arthropoda) characterised by having terminal claws on legs.

Echinonera. Old name for kinorhyncha (Aschelminthes). *Echinodermata:* phylum of animals with rough skin bec:mse of presence of spines. plates, ossicles etc., strictly marine and none parasitic - star-fishes. sea-stars, feather-stars. brittle-stars and sea-lillies.

Echinoida. Order of Echinacea (Echinodermata) having smooth test (oblong or spherical) and imperforate noncrenulate tubercles.

Echinoplutues. Bilaterally symmetrical larvae of sea urchins. *Echinostome*. carcariae larvae of flukes (*Trematoda*: Platyhelminthes) having collar of spines around oral sucker.

Echinozoa. Subphylum of free-living ecl}.inoderms having spherical body with meridional symmetry and no appendages.

Echiurida. (Echiura) small phylum of worm-like organisms having bilaterally symmetrical, unsegmented, sac-like or sausage-shaped body with anterior detachable prostomium.

Echiuroidea. Schizocoelomate animals. worm-like characterised by hind end spines - *Bonellia* species.

Echolocation. Auditory feedback mechanism used by bats and certain other animals (dolphins. propoises etc.) to find their way (avoiding obstacles) and food etc.

Ecological Niche. Limited ecosystem of a particular group of organisms. Many overlapping and interacting ecological niches make up an ecosystem.

Eclosion. Act or process adapted by insect for coming out of its egg. *Ecologist:* person having sound knowledge of ecology.

Ecology. Science of interrelationships of organisms in and to their environment.

Ecosite. Microparasite to which its host is immune under normal environmental conditions.

Ecosystem. Conglomerate of all life in a given area plus that life's interactions with the non-living portions of the area. Composed of biotic and abiotic components.

Ecotype. Animal or plant species locally adapted to particular environment or habitat.

Ectadenia. Assessory glands that arise from ectodermal layer of male ejaculatory duct.

Ectoblast. Outer cell layer of gastrula which gives rise to epidermis, sense organs and the nervous system.

Ectoderm Ectodermal. (outermost of the three cell layers of embryo; others being endoderm (innermost) and mesoderm (middle one).

Ectogenesis. Development outside the body or organism. *Ectogenous*. capable of living independently or outside the body of host.

Ectohormone. Pheromone; a substance secreted externally by an organism to influence the behaviour of other organism; an ectocrine gland's secretion substance.

Ectoparasite. Parasite of external body surface - tick, louse. bedbug etc. A parasite living outside the body of its host. Opposite of endoparasite.

Ectopic Pacemaker. Pacemaker situated outside the area where it is normally found.

Ectoplasm. Clear viscid outer layer of cytoplasm in animal cell. ECTOPROCTA: coelomate. sessile animals with 'U' shaped intestine, mouth surrounded by lophophore and anal opening outside the lophophore.

Ectosarc. Structure in protozoan composed of ectoplasm. *Ectosome*: outer, cortical layer of sponge

Ectotherm. Individual which obtains heat directly from environment. Vast majority of animals except birds and mammals are ectotherms. Compare with endotherms.

Edema. Accumulation of interstitial fluid in abnormally large amounts: caused by increased filtration pressure, decreased osmotic pressure, gradien't across capillary, increased capillary permeability and inadequate lymph flow.

Edentata. Order of ant-eaters (*Eutheria: Chordata*): mammals having no teeth, only poorly developed molars, long snout and extra articulation between lumber Vertebrae - armandilloes, sloths and ant-eaters.

Edentate. Ant-eater: any species belonging to mammal ion order Edentata.

Eel. Fish having snake-like body, naked skin and without pelvic fins. The dorsal caudal and ventral fins are confluent. Catadromous *i.e.,* they are found in fresh-water and go to sea for spawning -*Anguilla* species.

Efferent Blood Vessel. Blood vessel carrying blood out of a structure like gill or glomerulus of kidney.

Efferent Nerve. Nerve conveying impulses outward or away from brain nerve/spinal cord. Opposite to afferent nerve.

Egest. Defaecation: pass out or discharge undigested food/ residue from alimentary canal.

Egesta. Excreta: waste material thrown out from the body.

Egg. Non-motile female gemete. Term used generally for shelled structure containing yolk, albumen and 2i. cellular structure called nucleus/germinal disc.

Eating Snakes. Group of non-poisonous snakes (*Opidia: Reptilia*) found in India and Africa that feed exclusively on avian eggs: move from tree to tree in search of birds' nests, swallow whole egg which is then broken in the throat. its contents swallowed but pieces of shell thrown out through the mouth.

Egret. Group of white birds (Ciconiformes: Aves), live in marshy places around lakes and feed in shallow waters on fishes, frogs etc.

Egta. (*Ethylene Glycol-bis* (*β Aminoethylether*) N. N'*Tetra Acetic Acid.*): calcium chelating agent.

Eider. Sea-duck of northern hemisphere. *Someteria* spp. (An serif ormes: Aves) valued for its fine down.

Eimer Tina. Suborder of Eucoccida (Coccida: Telosporea) in which microgametocytes produce large number of microgametes.

Ejaculation. Discharge of semen out of urethra. involves emission and ejaculation proper (propulsion of semen out of urethra at the time of orgasm). Touch receptors at the glans of penis form afferent pathways and responses reach spinal cord through internal pudenal nerves. Emission is sympathetic response integrated in upper lumber segment of spinal cord and affected by contraction of smooth muscles of vasa deferentia and seminal vesicles in response to stimuli in hypogastric nerves. Semen is propelled out of urethra by contraction of bulbocavernosus muscle (a skeletal muscle). Spinal reflex centres for this part of reflex are in upper sacral and lowest lumber segments of spinal cord and motor pathways traverse the 1- 3 sacral root and internal pudenal nerves.

Eland. Large, heavy-built Mrican antelope. *Taurotragux oryx,* grey with white stripes, both sexes having spirally twisted horns.

Elaps. Poisonous coral-snake (Ophidia: Reptilia) of America.

Elastase. Enzyme found in succus enteric us (intestinal juice) that acts upon elastin and digests it.

Elastic Catilage. Cartilage in which yellow and elastic fibres are found in matrix. It is found in the pinna. epiglottis and trachial rings.

Elastin. Substance of the elastic fibres, a scleroprotein which is insoluable upon boiling but dissolves in proteolytic enzymes. Elastin fibres are found ir wall of large arteries and between vertebrae (intervertebral discs of mammals).

Elbow. Outer part of joint between fore- and upper arm.

Electric Potential. Electrostatic presure analogous to water pressure. A potential difference across a resistance is required for flow of current.

Electrocardiogram (ECG). Record of electrical events associated with contraction of heart, obtained by placing electrodes on various parts of body. The basic principle involves recording deflections of a string galvanometer from base level and designated by letters P. Q. R. Sand T. The parts of the line between letters are called intervals. The P-wave is produced by activation of SA-node. The PQ interval represents the time when auricles are contracting. Q-wave is ascent of R-wave preceedes ventricular contraction and show the spreading of contraction from AV-node through bundle of His and Purkinjie's system to ventricular muscles. Rest of QRS complex and most of S-T intervals occur as the ventricle is contracting. The T-wave is indicative of relaxation of ventricle.

Electrolyte. Substance that dissociates into ions in water solution and thus allows electric current to pass through solution.

Electromyography (EMG). Process of recording the electrical activity of muscle on cathode ray oscilloscope. It can be done in unanasthesised human beings by using small metal discs on skin overlying musclesas, the pick up electrodes or by using hypodermic needle electrodes.

Eleutherozoa. Subplylum of echinoderms composed of all extinct classes except crinoids (sea-lillies or sea-feathers).

Elephant. Largest living land animal (*Proboscidea: Mammalia*) grows upto four metres (shoulder height) characterised by elongation of nose and upper lip to form long prehensile trunk; unique dentition. in males two upper incisors are rootless and grow very long as tusks. yielding valuable ivory - African species *(Loxodont africana)* and Indian *(Elephas* maximus); about 13000 individuals in India.

Elevers. Miniature eels about 50-55 mm long.

Elk. Largest living deer. *Alces alces* (Cervidre: Mammalia). found in northern Eurasia. akin to 'moose of northern America but slightly smaller in size.

Elytron. Forewing of coleopterous insect that is thick and horny and covers the membranous hind wing.

Emasculate. Castrate: deprive of masculine proporties powers. *Embden, Meyerhof Pathway*: glycolysis (glycolytic pathway) named after two German scientists who worked out the details of the process in 1930.

Embioptera. order of little known insects (*Hexapoda*); mostly web-spinners, living in silken tunnels. mouthparts mandibulate; males Winged with gradual metamorphesis, females wingless and metamorphosis less pronounced–embiids.

Embolism. Blood clot forming tendency within a blood vessel. *Embolomerous vertebra*: vertebra with double centrum as found in most primitive amphibia; an anterior or hypo centrum believed to have been derived from basiventral and posterior pleurocentrum that have been formed from basiventrals and interventrals.

Embolus. Blood clot within a blood vessel. *Emboly*: invagination during gastrulation.

Embrasure. Space or cavity formed between two adjacent teeth. EMBRYO: multicellular structure developed from zygote and giving rise to new offspring generally in side female tissue or egg membrane.

Embryology. Science dealing with study of formation and development of embryc.

Emeriocytosis. Cell vomiting: cell membrane around a vacuole or secretion granule fuses with cell membrane and the region of fusion breaks down leaving contents of vacuole or secretion granule outside the cell membrane intact. Also known as exocytosis or reverse pinacytosis.

Emmetropia. Vision defect: when ciliary muscle is relaxed, parallel light rays striking optically normal eye. are brought to focus on the retina. As long as this relaxation is maintained, rays from the observer are brought to a focus behjnd retina and consequently the object appears blurred.

Emperor Dragonfly. Largest dragonfly (Anlsoptera: Odonata), about 10 cm long and wing span of 12 cm charactertsed by shape of hind wings, inner margins of which are rounded instead of angled. *Moth*: Fairly large moth *Satumta* species of silk-moth family Saturnidae (Lepidopte'ra: Hexapoda) charactertsed by large eye-spots on all four wings of both sexes, spins elaborate coarse brown silk cocoon which is of no commercial value.

Emphysema. Degenerative disease of lungs in which lungs lose their elasticity and their walls between alveoli break down so that the alveoli are replaced by air sacs. The physiologic ally dead space is greatly increased and because of inadequate and uneven alveolar ventilation and perofusion of underventilated alveoli, severe hypoxia develops. Inspiration and expiration are laboured and breathing is greatly increased.

Emulsion. Fluid formed by suspension of an oily or resinous liquid in another liquid.

Enamel. Outer hard layer of tooth that covers and protects the softer dentine within.

Encephalin. Two pentapeptides identified as natural pain killers of body by binding to opiate receptors. These are produced in brain. Two types are recognised - Met-encephalin and Leu-encephalin. Suffix Met- and Leu- correspond to 'C' terminal amino-acids methionine and leucine respectively.

Encephalisation. Relatively greater role of cerebral cortex in various functions. An important phenomenon in species in which cerebral cortex is highly developed.

Encephalitis. Inflamation of brain; usually caused by virus, or microbes: may be secondarily due to general infections such as that of measles, pneumonia etc.

Endergonic. Process or reaction characterised by concommitant absorption of energy *e.g.,* protein synthesis. Opposite to exergonic.

Endocardium. Innermost layer of heart derived from mesenchymal cells between endoderm of gut and splanchnic visceral mesoderm. Over this myocardium and epicardium are fonned.

Endocrine Glands. Glands without ducts. Their secretion is called hormone and is earned by blood, number of such glands are found in body like pituitary, thyroid. parathyroids, adrenals, interstitial cells of gonads, islets of Langerhans ana certain glands in alimentary canal.

Endocytosis. Engulfing of substances by cell because of membrane invagination proximal to particle and subsequent clossure of membrane over the particle (see phagocytosis).

Endoderm. Innermost of the three primary cell layers of embryo; others being ectoderm (outermost) and mesoderm (middle one).

Endogenous. Some substance/secretion corning from within the body.

Endogenous Insulin. Insulin secreted inside the body. it enters the portal vein so that liver is normally exposed to concentrations of insulin that are 3-10 times greater than those in peripheral tissues. Liver binds about half.a dose of insulin injected into the portal vein but only 25% of a peripherally injected dose.

Endolymph. Aquous liquid with high potassium concentration and found in scala media of cochlea of mammalian ear.

Endometrium. Uterine mucosa epithelium that lines the uterus. *Endoparasite*: parasite that lives within the body of its host *e.g., Entamoeba. histolyttca.* Taenia *solium, Ascaris* etc. Cf. ectoparasite.

Endoplasm. Cytoplasm that is surrounded by ectoplasm within the cell.

Endoplasmic Reticulu. Complicated pathway of channels and tubes in the cell that serves mainly to transport metabolites and is associated with conversion of chemical energy into mechanical work.

Enfotphind. Two large pep tides that are bound to opiate receptors and have analgesic properties/activities. One contains 17 amino-acid residues (a.-endorphin) and other contains additional-amino-acid residues (13-endorphin). Both contain amino acid sequence of met-enkaphalin. Arninoacid sequence found in 13-endorphin is in turn found in a polypeptide called 13-lipoprotein which has been isolated from anterior pituitary.

Endosome. Central mass consisting of chromatin 'material found in nucleus of certain protozoa like. *Amoeba, Euglena* etc.

Endotherm. Animal which produces heat within its body by oxidative metabolism and maintains constant body temperature like bird or mammal. Cf. endotherm.

End Plate. Traditional name for vertebrate neuromuscular junction where motor axon ramifies into fine terminal branches over a specialised system of folds in post synaptic membrane of a muscle cell.

Energy Pyramid. Means of describing energy relationships within ecosystem. Moving toward apex. amount of energy decreases. forming a sort of pyramid.

Energy Rich Phosphate Bond. Phosphate bond released by hydrolysis of ATP. Release of first bond liberates 3000 calories and the release of second bond 12000 calaries. Energy so released is used for various body activities.

Enterokinase. Intestinal enzyme in vertebrates that converts inactive trysinogen of pancreatic juice to active trypsin.

Enteroprocta. Pseudocoelomate animals. sessile with 'U' shaped intestine, mouth surrounded by a circle *0f* ciliated tentacles and anal opening within circle.

Entomology. Science dealing with study of insects. *Entomophagous*: organism that feeds on insects. *Entozoic*: living within an animal body.

Entropy. Measure of that portion of energy which is not available for work in a closed system; measure of molecular randomness. It increases with time in all irreversible processes

Enzyme. Organic substance produced by living cells that catalyses specific chemical changes like hydrolysis, oxidation, reduction, isomerisation, synthesis and analysis without undergoing any change itself.

Kinetics. Rate at which a reaction proceeds depends upon concentration of substrate, product and active enzyme.

Eohippus. Dawn horse. *Hyracotherium* species; Eocene (Tertiary) fossil mammal, about 80 ern long. said to be ancestor of modern horse. known for its speed and intelligence.

Eosinophilia. Malady characterised by increase in number of cosinophils from their normal percentage; generally associated with allergy reactions.

Eosinophils. Granulocytic white blood corpuscles of mammalian blood having bilobed nucleus and the two lobes being joined by bridge-like structure; Non-phagocytic and can be stained with acidic stains like eosin: normallv 1 to 4% in number; increase in number during infections and allergy.

Ephyra. (*Ephyrula*) free-living laxva of scyphozoan coelenterate like jellyfish *(Aurelia* species).

Epicardium. Visceral peritonium closely applied over the heart, derived from splanchnic mesoderm that surrounds the endocardial tube during the development of heart.

Epicuticle. During moulting those epidermal cells secreted under old skeleton (exoskeleton). Made of cuticulin. Layer of some chemical substance like wax or calcium present over the cuticle in some arthropods like insects and myriapods which prevents water-loss.

Epidemic. Widespread outbreak of disease at one time in one place. EPIDERMIS: outer cellular layer covering external surface of a meta£oan; derived from embryonic ectoderm and often secretes cuticle in some animals.

Epididymis. Coiled structure containing efficient tubules of mammalian testis. Embryonic mesonephros in mammals which after formation of metanephros joins into the sexvice of testis. Consists of three parts–caput epidiymis which receives various vasa efferentia, corpus epididymis which is the middle part and cauda epididymis which opens into vas deferns or spermatic duct.

Epiglottis. Flap of cartilagenous tissue in the larynx of mammals that closes the air passage to lungs during swallowing of food.

Epilepsy. Nexvous disorder in which patient loses· consciousness and falls down. A characteristic cry may often be heard. There rr.ay be mood disorders like sudden melancholy, anger and less frequently gaiety. Patient's memory is impaired.

Epimere. Mesodenn during embryonic development differentiates . into dorsal part nhich is called epimere. It differentiates into dermatome forming dermis and sclerotome forming vertebral column.

Epineural. Nervous tissue dorsal to ventral nerve cord in some arthropods.

Epiotic. Bone one of the five bones. fonning auditory capsule of fish. *Epiphallus*: sclerite in base of genital chamber in certain orthopterans.

Epiphysis. End of long bone-like femur: portion of bone having its own centre of ossification.

Epipolsina. Suborder of Clavaxinellida (Demospongiae: Porifera), sponges having radially arranged monactinal or diactinal megascleres.

Epistasis. Gene interaction in which one gene is able to mask the expression of other gene which has different locus and which is thus not its allele.

Episternum. Upper part of sternum in frog connected to omostemum joining with clavicles and epicoracoid. It is absent in toads.

Epistome. The plate between mouth and second antennre in crustaceans: in insects, sclerite between labrum and epicranium.

Epithalamus. One of the divisions of thalamus having connections to olfactory system.

Errantia. Order of polychaete annelids (Chaetopoda) with long linear body that is dorsoventrally flattened - *Aphrodite* species, *Eunice gigantea* etc.

Erythroblastosis Foetalis. Conditions in foetus where because of mother-father Rh factor' incompatibility, the infant's blood (Rh+) may clot when corne; in contact with mother's blooj (Rh-) . ather being Rh+.

Erythrocurorin. Haemoglobin present in the body of invertebrates like annelids, molluscs and few arthropods such as *Daphnia* and chironomous larvae.

Erthrocytes. RBCc of blood. Haemoglobin containing cells characteristic of all verte rates. Help to transport oxygen in the body. Rounded oval, biconcave or biconvex. mostly have a nucleus except in mammals, formed within lymph nodes and spleen in non-mammals and in the red bone marrow in mammals, life period is about 100 days in non-mammals like frog and 120 days in mammals. The number being 5 million/ml of blood of man. In women the number is less and in children the number is more. Process of their formation is called erythropoisis.

Erythropoisis. Process of RBCs formation. It is inhibited by a rise of RBCs in blood to supernormal value and stimulated by anaemia. Process is normally controlled by a circulating glycoprotein hormone called erythropoitin. Its steps can be summrnarised as below:
Uncommitted stern cells → Committed stern cells → Proerythroblasts → Early nomoblasts → Intermediate no mob lasts → Late nomoblasts → Reticulocytes → Erythrocytes.

Erythropoitin. Glycoprotein hormone circulating in the blood formed by a renal factor on a globulin in plasma. It promotes differentiation of committed stern cells (erythropoitin sensitive cells) into proeythroblasts.

Estrus Cycle. Mammalian egg production and fertilisability; reproductive cycle period in mammals when breeding occurs. Compare with mentrual cycle.

Ethologist. Person having sound knowledge of ethology.

Ethology. Science dealing with the study of animal behaviour based on comparisons. The observations of patterns and consideration of environmental adaptation.

Etiology. Assignment of cause of disease or science dealing with the study of causation.

Eucoccida. Order of Coccidia (Telosporea: Sporozoa) having alternate sexual and asexual phrases - parasitic protozoans.

Euechinoidea. Subclass of Echinoidea (Echinodermata) characterised by relative stability of ambulaca and tnterambulaca.

Eugenics. Science dealing with application of knowledge of heredity to improve human species. Term was coined by Sir Francis Galton in 1883. It incorporates two types of measures- one discouraging the reproduction in inferior or abnormal human beings (negative eugenics) and the other encouraging the production of superior human beings (positive eugenics).

Eurypterid. Extinct chelicrate arthropod of palaeozoic era. EUSTACIUAN TUBE: tube leading from middle ear to phrynx in higher verebrates; named after Italian physician Custachi.

Eutheria. Placentalia; group of mammals in which young ones are retained for consitlerable time in uterus of mother and nourished by placenta; includes all mammals except egg-laying ones and marsupials.

Euthyroid. Patients with abnormal thyroxine binding globulin level and do not show signs of hyper or hypothyoidtsm.

Eviscerate. Removal of intemal organs/viscera from an animal. *Evolution*: process whereby living beings are believed to have arisen from less advanced forms by series of gradual changes.

Eecitability. Property of altered membrane conductance (and often membrane potential) in response to stimulation.

Excitable Tissue. Tissue that can be excited by chemical, electrical or mechanical means and produces an action potential which is transmitted along its membrane. Includes muscles and nerves.

Excrement. Dung: unwanted matter discharged from animal body. *Excreta*: matter discharged from animal body: faeces and urine. *Excretion*: removal of metabolic wastes also the substance discharged. 'In strict sense removal of nitrogenous wastes in the body.

Exhausation Atrophy. β-cells of islets of Langerhans respond to stimulation with hypertrophy like other endocrine glands: they become exhausted and stop secreting when stimulation is marked or prolonged. This is called exhausation atrophy if stimulation stops soon after this they can recover but if stimulation is continued, the cells eventually die and disappear.

Extensor. Muscle that straightens out or extends a part of body as foot. Opposite to flexor.

Exterorecetors. Sense organs that respond to external stimuli receptors for touch, smell, taste, sight and hearing.

Extertional Rhabdomyolysis. In potassium depleted individuals muscles. blood flow increases to a lesser degree during exercise and there is a greater tendency for severe disintigration of muscles which refers to extertional rhabdomyolysis.

Embryonic Membranes. Cellular membranes found outside the body of embryo during development of reptiles, birds, mammals (amniotes). These are amnion, chorion, allantois and yolk sac. Amnion and chorion have protective role and also provide optimum moisture for developing embryo; allantois performs respiration and excretion while yolk sac has nutritive role.

Exumberella. Convex, dorsal or aboral surface in some animals like jelly-fish or medusa. Opposite to oral surface/subumbrellar surface.

Eye. Photoreceptor in the body having three layered cup and a lens. Separated from exterior by conjunctiva and enabling the animal to have vision.

Eye Spot. Simple eye or ocellus. organ of vision in many invertebrates.

Eyeball. Entire globe of vertebrate eye without appendages. EYEBROW: hairy arch above the eye

Eyestalk. Movable peduncle bearing terminal eye, as in decapod crustaceans.

F

Facet. External surface of an individual ommatidium: cavity depression in skeleton element meant for articulation.

Facilitated Diffusion. Downhill transmembrane diffusion aided by a carrier molecule that enhances the mobility of the diffusing substancein the membrane.

Faces. Undigested material passed out of food canal through anus FAINTING: syncope: sudden loss of consciousness due to cerebral ischemia Most important cause is hypotension due to sudden vasodialation of blood vessels and bradycardia (decreased heart rate), often associated with strong emotion. Attack short-lived and consciousness is restored in a few minutes. This is called vasovagal syncope. Pooling of blood in dependent parts of body on standing may cause postural syncope. Persons with orthostatic hypotension have miturition syncope. Cough syncope is when there is increase in intrathoracic pressure, during coughing, enough to block venous return. Effort syncope is fainting on exertion due to inability of cardiac output to meet increased tissue demand occurs in persons having aortic/pulmonary stenosis.

Falcon. Group of small birds of prey, *Falco* spp. (Falconidce: Aves) with long-pointed wings and long tail - gyrfalcon. kestrel. peregIne falcon etc.

Falconid. Family of long-winged. diurnal, predaceous birds-(Falconiformes : Aves) that includes falcons.

Falconiformes. Order of birds of prey (Cartnate: Aves) that hunt by day: have sharp, strong curved beak and powerful feet and claws – hawks, kites, vultures.

Fallow Deer. *Dama. dama* reddish-brown deer with white spots and white patch on rump; male (buck) carries broad poalmate antlers; herbivorous; found in Mediterranean region.

Family. Principal subdirision of order of animals/plants.

Fangs. Speciallsed grooved teeth over maxillae of poisonous snakes like, Cobra, Viper, kraits etc., through which a pair of poison glands open.

Faraday Unit. Unit of electrical capacitance (F) expressed as the equivalent charge of a molecule of electron equal to $9.649 \times 10.^4$ coulombs per molecule of electron.

Fascia. Band connective tissue that covers and supports or binds parts together.

Fasciculation. Jerky visible contractions of muscle fibres due to pathogenic or pathological discharge of spinal motor neurons.

Fasciculi. Pathway structure present in brain stem to which second order neurons pass down the spinal cord from vestibular nudei in the vestibulospinal tracts and ascend through these to motor nuclei of cranial nerves concerned with control of eye moveulent.

Fatigue. Build up of waste products e.g., CO_2 and lactic acid preventing muscular contraction in a muscular tissue (ability of muscle to contract is reduced or obliterated).

Fat Metabolism. Fats after digestion are converted to glycerol and fatty acids; these substances are metabolised in two ways:

Fats. Organic compounds having - C. Hand 0; esters of fatty acids and alcohol. If the alcohol is glycerol they are natural fats but if it is of higher grade then these are called waxes. Yield double the amount of energy (9.3 ccls/gm) as compared to equivalent weight carbohydrate/protein.

Fauna. Animal life of a given pertod or region.

Feather. Structures found on skin of birds. These make body light and insulated from heat. cold, electrtcity; etc., help in balancing the body when attached to tail; help in flight when attached to Wings; make general contour of bird. Many types - contour feathers, down feathers. floplumes and bristles.

Fecundity. Innate potential reproductive capacity of an individual organism.

Feed-back. Return of output to the input part of a system. In negative feed-back sign of output is inverted before being fed back to the input so as to stabilise the output. In positive feed-back the output is unstable because it is returned to the input without a sign of inversion and thus becomes self-reinforcing or regenerative.

Feed Back Control. Control through return of the output to the input part of the system. Mechanism by which activity of a system is governed on the basis of information derived from the output system's activity. A system in which accumulation of product leads to a decrease in its rate of production or deficiency of product leads to an increase in its rate of production.

Femur. Bone of hind limb present in thigh. It is articulated anteriorly with acetabulum of pelvic girdle and with tibia and fibula posteriorly: third segment of an insect leg.

Fenestra Ovalis. Oval gap in bony labyrtnth of mammals which connects membranous labyrinth with middle ear; covered with vibratile membrane and over this is fitted stapes.

Fercula. Wish-bone or marry-thought bone formed by fusion of clavicles and interclavicle in the birds. A fork-shaped structure articulating anteriorly between two pectoral girdles and postertorly articulated with sternum.

Fermentation. Enzymatic decomposition; anaerobic transformation of nutrients without net oxidation of electron transfer.

Ferritin. Large protein molecule. opaque to electrons used as marker in electron microscopy: normally present in spleen as storage protein for iron.

Fertile Period. Period of 72 hours (14. 15. 16th day) during which ovum is released into the upper part of oviduct and fertilisation is possible. The hormonal level during this period is at peak.

Fertilisation. Union of mature ovum and mature sperm to form zygote.

Fever. Oldest and most universally known hall-mark of disease during which the thermoregulatory mechanisms behave as if they were adjusted to maintain high body temperature level than *3TC*. A shivering. often in fever is due to activation of temperature rising mechanism. Drugs stimulate thermoregulatory centres to relieve it by reducing set point back to normal.

Fibrin. Blood protein and major component of blood clots in the vertebrates.

Fibrillations. Soluable form of fibrin which pn coming on in contract with thrombin forms fibrin. a component of blood clot.

Fibrinolytic System. System in blood that limits clotting and decomposes clot to fibrinogen degradation products. Active component of the system being fibrolysin or plasmin. This substance lyses fibrin and fibrinogen with the production of fibrinogen degradation proeucts that inhibit thrombin. Plasmin is formed from plasminogen. There is an activator circulating normally asproactivator. The proacuvator is activated by proteolytic fragment of active foactor-XII formed by the action of plasmin on active factor-XII. This positive feedback of plasmin on its own formation gets limited by available store of factor-XII and by various other limiting factors.

Fibroblast. Connective tissue cell that can differentiate into chondroblasts. collagenoblasts and osteoblasts.

Fibrous Joint. Joint without a cavity; bones are directly opposed. *Fibulia*: narrower to two bones in vertebrates' hind limb between knee and ankle. In frog and related animals, it is fused with tibia to form tibiofibula bone.

Fiddler. Small crab. *Uca* species (Decapoda : Crustaces); male having one of the two pincers (claws) greatly enlarged.

Fish. Long-tailed. shark-like fish (Selachti; Ptscer)

Fightingfish. Small b1"ightly coloured fish (Pisces), bred for centrures in Thailand (Siam) for fighting (as sport).

Filial Generations (Fiand F_2): first and second filial generations with respect to successive generations following cross-breeding (the hybrids produced by cross between original parents constitute F} generations and the products formed by selling the F_1 hybrids constitute F_2 generation}.

Filoplume. Minute hair-like feather with few barbs at the tip. *Filopodium* : filamentous pseudopodium of sarcodine protozoan.

Filter-Feeder. Microphagous aquatic organism that has filtering mechanism for catching food particles found suspended in water.

Filtration. Process by which fluid is forced through membrane or the other barrier due to a difterence in hydrostatic pressure on the two sides. Amount of filtrate is proportionate to pressure difference and surface area of membrane. Molecules of smaller size than the pores of membrane pass with the filtrate while larger molecules are detained.

Filum Terminale. Thread-like terminal part of spinal cord. Non-nervous. In mammals this part terminates in lumber region even if there is a tail.

Fin. Locomotor or starting structure and extension of body in aquatic animals. These may be paired/unpaired. The paired fins include pectoral and pelvic fins while 'ventral anal and cudal (tail) fins are unpalred ones.

Final Common Pathway. Motor neurons that supply extra-fusal fibres In skeletal muscles are dimerent side of reflex arch thus all neural Influences affecting muscle contraction are ultimately funnelled through them and for this reason they are said to fonn final common pathway.

Finger. Any of the four digits of hand except thumb. *Printing*: procedure used to separate component parts of protein following enzyme cleavage of protein into fragments and their separation by a combination of electrophoresis and paper chromatography.

Finnan. Haddock fish *(Melanogrammus* species) cured with smoke of green wood; relished as delicacy.

Firefly. Glow-worm a beetle (Lampyrldre : Coleoptera) that emits light (in flushes) by night.

Firing Level. First manifestation of approaching Impulse is beginning of depolarlsation. After an initial 15 mv depolarlsation Its rate Increases. The firing level is the point at which this range in rate occurs.

Fish. Cold blooded aquatic animal that breathes by means of gills and swims by means of fins.

Fishery. Place for breeding and harvesting fishes.

Fission. Asexual method of reproduction by division into two or. more parts, approximately equal in size.

Fissipedia. Suborder of Carnivora (Mammalia': Chorclata that indudes terrestrial animals - bear, dog, lion, wolf etc. project of tuft of cilia. The cytoplasm of cell absorbs wastes and these are passed into duct by tuft of cilia so that they are passed to excretory canal.

Flare. Redness spreading out from injury and is due the arteriolar dialation. The pain In this area is due to chemical agents liberated by antidromic impulses in sensory nerve fibres. The agent is a proteolytic enzyme and causes liberation of vasodialator kinins. The kinins lower threshold of pain 1lbres In the area and they may themselves produce patn.

Flavin. Ayellow pigment which combines with proteins and transfers electrons to cytochromes.

Flavoproteins. Proteins combined with flavin prostheilc group important as intermediate carriers of electrons between the dehydrogenases and cytochromes in respiratory chain.

Flesh. Soft substance of animal body between bones and skin. FLEXOR: muscle whose function is to bend or flex a joint, therehy decreasing the angle of two component parts, Opposite to extensor

Fligh or Fight. On being threatened the first attempt of antmalls to flee, but on being cornered in such situation, it attempts to counteract. These responses are called flight or fight respectively.

Flocculus. Lateral portions of cerebellum called flocculonodular lobe. Its connections are essentially all vestibular. Animals in which It has been damaged or destroyed, walk in staggering fashion on the broad base. They tend to fall and are reluctant to move without a support.

Flounder. Common name for variety of edible flat-fishes (Pleuronectiformes : Pisces) of shallow sells around Atlantic ocean - *Platiclithys flesus* etc.

Flock. Antmals of one kind - birds, cattle etc.

Fluorescence. Property of emitting light upon molecular excitation by an incident light. The emitted light is always less energetic (of short wave length) than the light producing excitation.

Fluorescent Strains. Those strains that can combine with chromosomes and render them lumtnous as a result of their ability to emit light some of which is absorbed.

Flux. Rate if flow of matter or energy across a unit area.

Fly. Any dipterous insect (Hexapoda) characterised by having only one pair (forewings) of membranous Wings; hind wings reduced to halteres.

Flying Fox. Common name for flying mammal *Pteropus medius* (large frugivorous bat). The animal has rufous head and nape, black neck, golden/yellow shoulders. dark brown back. dark chin and rest of body beneath rusty brown. Ears small, very short tail and wing-span 1.5 meters.

Frog. Amplhiblan *Racophorous* spp. (Anura) having very long and webbed toes that enable them to long leaps from tree to tree conspicuously coloured: Indian species *(R. paradalusJ* has black body, shlnir.g green limbs, webs black with yellow tinge and inner toes yellow. R. *arboreus* Is found In Japan and R. *leucomystax* in China.

Foetal Membranes. Extra embryonic membranes. *Foetus*: advanced stage of mammalian embryo.

Foliaceous. Leaf-like appendage which may also be multilobed. *Foliate Papillae*: broad. leaf-like papillae lying near base of tongue in parallel rows. Bear taste buds on their sides. Not present in man.

Folic Acid. Water soluable vitamin (member of vitamin B complex group) essential coenzyme for 'one carbon' transfer involved in methylation reactions. Found in leafy green vegetables. Deficiency causes sprucc, anaemia dt:.

Follicle. Cellular sac or covering. Term generally applied for a mass of cells surrounding developing ovum (Graaffian'follicle); mass of cells surrounding hair shaft (hair follicle).

Follicular Phase. Part of estrus and menstrual cycles that is characterised by formation of and secretion by Graafian follicle.

Foramen. Opening in a bone, membrane through which blood vessesls and/or nerves usually run.

Fore Brain. (Prosencephalon): Anterior of the three divisions market out in the embry Ouic brain of vertebrates by constrictious. Gives rise during development to cesebral hem ispheres, eyes, hypothalamus, etc.

Foraminifera. Sarcodine protozoa having fragile calcareous shells with pores through which pseudopodia come out.

Form (Morphotype). Distinguishable sympatric and synchronic interbreeding population of single species.

Formicivorous. Feeding on ants - some birds and mammals. *Farmicoidea*: monofamillal superfamily of Apocrita (Hymenoptera: Hexapodal - black ants, red ants etc.

Fornix. Band of longitudinal fibre below the corpus callosum in mammalian brain. The two strands of fibres called taeniae hippocampi form the fornix. The ridges are separated both in front and behind by main body of fornix forming anterior or posterior pillars. Through posterior pillars, the splanium is united. The anterior pillar passes ventrally connecting hippocampal lobe with hypothalamus and mamillary body.

Forward Failure. Cardiac output inadequate to perfuse muscles. specially failure of output to rise with the excercise. Such persons cannot tolerate excercise.

Fossils. Remains or other indications (casts, imprints, impregnations etc). of animals/plants which existed in the remote past ages.

Founder Effect. Genetic drift where new population is founded by relatively few individuals. hence their limited gene pool is the basis for all future generations.

Fourth Sound. During each cardiac cycle, besides normal first and second sounds, often third and fourth sounds are also heard in the stethescope. It can be heard before first sound when atrial pressure is high or ventricle is stiff in conditions such as hypertrophy. It is due to rapid ventricular filling and is rarely heard in normal adults.

Fovea. Small pit or depression in retina of some vertebrates.

Fowl. Bird of poultly: common name for both male and female.

Fox. Small burrowing mammal. *Vulpes benga!ensts,* and allied species (Canidre : Carnivora), reddish-grey in colour about 60 cm long with bushy tail another 40 cm; pointed snout, large erect ears and short legs: hunts after sunset preying upon rabbits, rodents, lizards etc.

Frame Shift Mutation. Genetic mutation where one. two (three+one)n. (three+two)n base pairs are delected or added. Leads to the formation of disturbed code and as a result different amino-acids may be coded for. May also result In premature synthesis termination.

Free Energy. Energy available for work in a system at a given temperature and pressure.

Free Martin. Female member of unlike sered twins in cattle and occasion in other unqeelates. Sterile, and partially cowerted hermaphrodite condition by some influence of its twin brother.

Hopper (Spittle Bug). Small greenish-yellow jumping homopteous insect (Cercopldre : Cicadoidea) found in grasses. pines etc .. nymphs produce froth or cuckoo-spit and live therein.

Froggery. Place where frogs abound; or are kept for breeding and reciring purpose.

Frons. Anterior portion of the fore head of an insect.

Frontal. Parallel to main body axis and at right angles to sagittal plane.

Frontoparietal. Wall of long flat bone forming roof of cranium in the vertebrates like frog.

Fructose. Keto-hexose sugar found (mostly In free state) mainly in plant tissue; component of sucrose. In solution forms furanose.

Functional Group. Commonly found group of atoms showing characteristic reactions.

Furrowing. The cytokinesis characteristic of animals cells. A constriction of the cytoplasm into two portions.

G

Gaba. (*Gamma amino-butyric acid*): inhibitoty transmitter identified in crustacean motor synapses and in the vertebrate nervous system.

Gagging. Something thrust into mouth to prevent speech or to hold it open for operation. This is a reflex response integrated in medulla oblongata.

Galactorrhea. Dopamine like aldopa inhibts prolactin secretion in mammals and during its disturbance there is abnormal secretion. The drugs increases prolactin formation and stimulate this process; and there is' excessive milk secretion.

Galactose. Constituent of milk, sugar lactose and formed by action of lactase enzyme on it. In the liver it is metabolised by its conversion to glycogen or to galactosamtn which is an essential constituent of nervours system.

Galactosemia. Inborn error of metabolism in which there is congenital deflciency of phosphogalactose uridyl'transferase responsible for reaction between galactose I-phosphate and uridine diphosppoglucose so that ingested galactose accumulates in the blood.

Gall Bladder. Bladder-like structure found in the liver. It receives a number of cystic ducts from liver lobes and give rise to a single duct mostly and rarely two bile ducts. The bile ducts bring stored bile to the duodenum.

Galliformeo. Order of domestic game-birds (Carinatae : Aves); terrestrial and grain-eating- partidges. peacocks. pheasants etc.

Gametocyte. Potential gametes of some merozoltes of malarial parasite *(Plasmodium)* which do not undergo further development as long as they remain in the man. Inactive and harmless to the man but become active when sucked by secondary host mosquito during its feeding operation on such a person.

Gametogenesis. Process of gamete formation: can be divided into two types- spermatogenesis which involves formation-of sperms by the testes and oogenesis that innvolves formation of ova by the ovaries .

Gamma-Motoneuros. Neurons orginating in the spinal cord and synapsing with intrafusal muscle fibre.

Gamma-Rays. Electromagnetic radiations of very short wave length 0.0 -12 cm) and have very high energy.

Gannet. Large, gregarious, fish-eating, sea-bird, *Bula bassana* (Sulid: Pelecaniformes); white with yellow head and black-tipped Wings; nesting in colonies on rocky coasts.

Gar. (*Garfish, Gar-Pike*) fresh-water fish (Belonl1dre: Osteichthyes) of North America. characterlsed by long-snout; feed on small-fishes-short-nose gar. *Leptdosteus platystomus.* long-nose gar L. *osseus.*

Garefowl. Extinct great-auk. a, sea-bird *Pingutnus* tmpennts (Alcid : Aves) of North Atlantic ocean.

Garget. Inflamation of udder (cow. ewe or goat).

Garnet. Gumard: fish with large angular head and three finger-like walking rays in front of pectoral fish.

Gasterophilid. Family of horse-botflies (Dlptera : Hexapoda) stout-flies with bare arista and spotted wings - *Gasterophilus intestinaUs.*

Gastralia. Abdorninal ribs or ventral ribs present in some extinct amphibia. liVing fossil Uzard *Sphenodon* and the crocodiles;

'V' shaped dermal bones in the ventral region of the abdomen. Dermai bones.

Gastric Juice. Fluid secreted by cells of gastric epithelium. It is a mixture of water; enzymes like pepsin. renin etc. and salts. It is digestive in function.

Gastrin. Protein hormone that is liberated by gastric cells and pyloric glands and induces the secretion of gastric juice by the stomach.

Gastrocentrous Vertebra. Vertebra found in reptiles and mammals in which pleurocentrum enlarges to form centrum while hypocentrum is reduced to a small piece (as in reptiles) or it is changed to form an intervertebral'disc (an hi mammals).

Gastrocoel. Primitive digestive cavity of metazoan embryo formed by gastrulation.

Gastrology. Scientific knowledge of diseases of stomach. GASTRONEMIUS: large muscle on posterior side of lower leg of a vertebrate.

Gastropoda. Class of mollusca. having foot attached below stomach looking like modification of the later-snails and slugs.

Gastrovascular. Cavity /space serving functions of both digestion and circulation. coelentro n of *Hydra.*

Gastrula. Stages in development in which the embryo usually consists of two germ layers (ectodrem+endoderm) with a cavity surrounded by endoderm (Archenteron).

Gastrulation. Process of gastrula formation during the development of an organism; invagination of blastula.

Gazelle. Small, graceful, fast running antelope, *Gazella bennetti* (Artiodactyla: Mammalia) found commonly in Asia and Africa.

Gecko. Small. thick-bodied. dull-coloured lizard, *Gecko* spp. (GechonidlE : Lacertilia) haVing adhesive toes and vertebrae concave at both ends; found in warm climate, distinguished by their peculiar cry.

Gemsbok. Large antelope *Oryx gazella* of South-western Africa; having long. straight, pointed horns.

Gena. Lateral portion of insecs head : feathered portion of mandible (lower jaw) of bird.

Gender. Sex: three fonns – feminine, masculine and neuter. *Genealogise*: to trace one's pedigree.

Genealogy. History of descent or pedigree of particular Individual or family.

Gene Flow. Flow of gene pool of alleles from another population through interbreeding that changes the original allele frequencies.

Generation. Group of individuals of same species born around same time.

Genes. Units of heredity which are transmitted in chromosome and which by interaction with Internal or external environmental control the development of a trait/traits; the heredity determiners.

Genetics. Science dealing with study of laws and machanlsms related with transmission of characters (traits) from parent to offsprings: science of inheritence.

Geniculate Body. Projection of geneculate nucleus that relays incoming sensory information to the cortex named for its knee-like shape in cross section.

Genitals. (*Genitalia*): reproductive organs of male or famale Individuals. Internal genitalla refers to reproductive organs like testes, vas deferens etc. in male and ovaries, oviducts, uterus etc. In female. External genitalia refers to copulatory organs like penis, scrotum in male and vulva. opening of bulbourethral chamber, clitoris etc. In female.

Germplasm. Material basis of heredity. Term given to hereditory substance in germ cells by Weismann.

Gestation. Period between fertilisation and the birth of an eutherian mammal. During this period the entire development occurs. It varies in different mammals; gestation periods of some common mammals are:

Bitch, cat	-	60 days/about 8-9 weeks.
Cow	-	280 days/about 40 weeks
Elephant	-	600 days/about 85 weeks
Ewe	-	147 days/about 21 weeks
Guinea pig	-	68 days/about 9-10 weeks.
Mare	-	336 days/about 48 weeks
Monkey	-	168 days/about 25 weeks
Mouse, rat	-	21 days/about 3 weeks.
Opposum	-	13 days/about 2 weeks.
Rabbit	-	32 days/about 4-5 weeks.
Sow	-	112 days/about 16 weeks
Woman	-	270 days/about 39 weeks

GFR. (*Glomerular Filtration Rate*): Amount of total glomerular filtrate produced per minute by all the nephrons of both the kidneys. Equals to clearance of a freely filtered and non-reabsorbed substance such as insulin.

GH. (*Growth Hormone*): Protein hormone secreted by anterior lobe of pituitary and stimulates growth by action upon thyroid gland and other body functions like protein. carbohydrate and fat metabolisms. It is also called somatotropin.

Giant Panda. Black and White bear-like mafi!mal *(Ailuropus melanoleucusl* about 1.8 metres high; found in China and Tibet.

Gigantism. Excessive growth of body owing to higher jhypersecreUon of growth hormone by anterior lobe of piiuitary since birth. Compare with dwarfism.

Gila Monster. Bearded-lizard: any of two species of *Helodemla* (Helo dennatldre: Squamata); noctumal, poisonous, prey upon mice and small lizards.

Gizzard. Muscular part of the digestivc tract as in earthworms. cockroaches and birds; useful for grinding the ingesteo food materials in the want of a grinding apparatus in the buccal cativy. .

Gland. One or many associated cells that secrete or excrete one or more specialised substances. These are of several types: According to number of cells. glands may be unicellular (characteristic of many amphibia) or multicellular. The Illultic'eliular glands may be simple. unbranched tubular as sweat glands; compound branched tubular as mammary glands of prototherians: simple but branched flask-shaped glands as present along the margins of eyelids; simple flask-shaped alveolar) as the mucous glands of frog's skin: branched alveolar glands also called compound saccular glands as mammary glands of other mammals. According to mode of secretion. glands may be merocrine where cells discharge their products causing much damage to cells and the cells have to undergo a period of rest *e.g.,* mucous glands of amphibain skin; holocrine,. in which the cells of gland. when full \\1th capacity die and products are discharged from epithelial walls *e.g.,* sebaceous glands: apocrire in which the products collect at the outer part of cells *e.g.;* sweat glands of mammals; eccrine which are similar to apocrine glands but the cytoplasm is broken down before products are discharged *e.g.,* sweat glands.

Adrenal: two small yellowish bodies. lying just above each kidney secrete adrenalin that is directly absorbed into blood-stream. *Anal*: any gland in or around anus. as founed in certain chordate animals odoriferous anal· glands of snakes. *Antebrachial*: odoriferous gland in carpal region of certain primates (Mammalia) - lemurs etc. *Brachial*: odoriferous gland in proximal region of forelimbs (arms) as in males of some primates like lemurs etc. *Buccal*: any gland opening in or 'lear mouth (buccal cavity) -salivary glands. *Ceruminous*: wax-glands of ear. *Crop*: found in crop of some birds (both sexes of pigeon). These secrete milky nutrient fluid for feeding young ones. *Cutaneous*: common name for various skin-glands, such as sebaceous glands. sweat glands etc.

Endocrine: ductless glands; their secretions (hormones) being absorbed directly into blood or lymph vessels. *Exocrine*: glands having duct to cany off their secretions.

Glaucoma. Increased intraocular pressure caused due to blockage of canal of Schlemn through which old eye humors (aquous and vitreous humors) pass out. Characterised by intense pain in the eye-ball, mayor may not be accompanied by vision loss.

Glission's Capsule. Liver lobules in some mammals *e.g.*, pig are distinctly separated from one another by connective tissue and these interlobular septa are referred to as Glisslon's capsules.

Globe-fish. Common name for various tropical fishes belonging to families Dlodontldre and Tetrodontldre, having ability to inflate themselves into a ball.

Globulins. Simple proteins which coagulatt" on heating. Insoluable in dilute solutions of salts; found in egg white, blood. seeds etc. These may have antibody function;

Glochidwm. Bivalve larva. of fresh-water mussel bearing pincers. Found in *Lamellidens.*

Glomerulcus. Called mass of blood capillaries arising from afferent renal and efferent renal arteioles, found inside the Bowman's Capsule and help in the process of ultraftltration.

Glomous Tissue. Tissue present in carotid/aortic body. Contains islands of two types of cells suroundcd by sinusoidal vessls. The cells are named type-I and type-II. Type-II cells are glial cells asurrounded by type-I cells. Unmyelinated endings of glossopharyngeal neive fibres are found, at intervals. between type-I and II cells.

Glossa. Tongue or tongue-like structure - median projections of labiusm - as in insects (Hexapoda).

Glucaon. Protein hormone released by the a-cells of islets of Langerhans. It secretion is induced by low blood sugar or by growth hormone. It stimultaes glycogenolysis *i.e.* brings about coversion of glycoi;en to glucose to meet out the problem of hypoglycemia.

Glucocorticoids. Steroids synthesised in the adrenal gland cortex with a wide ranging matabolic activies. Included are cotisone. cotisol. corticosterone and II-deodxycorticosterone. These regulate mainly the carbohydrate metabolism in the body.

Glucogenic. Glucose forming process *e.g.*, amino-acids that metabollse like carbohydrates after deamination.

Gluconeogenesis. Synthesis of carbohydrates from non-carbohydrate sources like fatty acids and amino-acids.

Glucostasis. Regulation of blood sugar in the body so that its normal level is always maintained. It is done by a combined effect to two hormones namely insulin and glucagon which .are responsible for glycogenesis and glycogenolysis respectively. The process is influenced by PSH (Pancreas stimulating hormone) from pituitary secreted under the control of PSIi-RF (pancreas stimulating hormone releasing factor) from hypothalamus.

Glutamic Acid. One of the products formed during Kreb's cycle. It is also formed by transfer of NH_2 group over u-ketogluteric acid. It can depolarise mammalian neurons when placed on their membranes.

Glycerokinase. Enzyme which is responsible for phosphorylation of glycerol and results in the formation of 3-phosphogly ceraldehyde during the metabolism of glycerol. The substance so formed enters the glycolytic pathway. It is an instance of intermediary metabolism.

Glycholic Acid. Cholic acid formed in the liver by metabolism of cholesterol and conjugated within the liver to form sodium and potassium glycholates.

Glycine. Amino-acid that can be synthesised in the body provided other nutrient elements along with nitrogen and carbon compounds continue to be available. It is mediator responsible for direct inhibition in spinal cord. It is found in neurons and in almost all living cells.

Glycogen. Homopolysaccharide formed within the liver from excess of glucose, fructose and galactose etc., It is stored in the liver, muscles and other tissue as reserve 10od. The process of its formation is called glycogensis. When body glucose level is lowered the substance is converted to form glucose through process of glycogenolysis. Its synthesis is controlled by hormone insulin and break-down is controlled by glucagon.

Glycogenesis. Process by which excess glucose in the body is converted to glycogen. Occurs within the liver. The process is regulated by hormone insulin and helpful in preventing hyperglycemia in body. In event of the disturbance in this process the sugar is seen passing into blood and urine causing a pathological condition known as diabetes mellitus.

Glycogenolysis. Conversion of glycogen to glucose when the body blood sugar falls below normal (as occurs during hunger). The process occurs within the liver and regulated by hormone glucagon. Both glycogenesis and this process help in maintaining glucostasis in the body.

Glycogen Synthetase. Enzyme that catalyses the formation of glycogen from glucose.

Glycolipid. Lipid conjugated with a carbohydrate molecule (galactose in most cases). Found generally in the nerve tissue and important constituent of it.

Glycolysis. Anaerobic respiration of carbohydrates, involving step-wise breakdown of glucose, starch and glycogen etc. to form pyruvic acid (a substance that is required to start Kreb's cycle). It takes place within the cytoplasm and outside the mitochondria.

Glycosuria. excretion of excessive amount of glucose in mine. *Gnasthostomata*: group of vertebrates having a pair of jaws. Most of the vertebrates belong to this subdivision *e.g.*, fishes, amphibia, reptles, birds and mammals. Opposite to agnathostomata.

GNU. Wild beest group of large ugly African antelopes with buffalo or cow like head, short horns and long hairy tall; shoulder height about one metre and weighs about two quintals.

Goat. Horned and bearded quadruped ruminant, *Capra Hicus* (Artiodactyla: Mammalia). strong-smelling, hardy animal; feeds on vegetable matter; domesticated and 'reared for milk and meat. *Antelope.* Group of hollow-homed ruminants in between goat and antelope such as *Nemorhaenus goral* of India. chamois of Europe and mountain-goat of North America. *Sucker*: nightjar, migratory bird, erronesusly believed to suck milk of goats.

Goblet Cells. Modified epithelial cells which secrete mucous over mucous membranes.

Goby. Small group of colourful marine, spiny-fish. *Gobius* spp. (Gobldre : Pisces). less than 10 cm long, fiattened head. large eyes and ventral fins united to form a sucker, with which they cling to rocks etc.

Goitre. Abnormal Increase in size of thyroid gland usually associated with dietary lack of the iodine.

Goitrin. Active antithyroid agent. Brasslca vegetables (family Cruciferae) particularly cabbage and turnip contain progoltrin and a substance which converts these corppounds into goltrln. It may cause goitre and Is goitrogen. Chemically the substance Is L-5-Vinyl 1, 2-thioxazolldone.

Golgi Bodies. Cytoplasmic organelles consisting of paired membranes and vacuoles. Can be stained with silver or osmium salts. Concerned and a number of cells arranged to form stratum germinativum and cumuls oophorous. In between these cell layers there is a cavity called antrum folliculi which is filled with a liquid called liquor folliculi. The whole complex helps in maturing the ovum. After maturation the ovum is released and then whatever is left in the follicle is called corpus luteum. The corpus luteum becomes a well developed endocrine gland if pregnancy has occur red and then it secretes various hormones which help in the maintenance of' pregnancy, birth of child etc. but regresses if pregnancy has failed.

Granded Response. One that increases as a function of the energy applied. A membrane response that is not all-or-none.

Granulated Endoplasmic Reticulum. See ergastoplasm. *Granulocytes*: WBCs having granulated cytoplasm. Three types can be recognised - eosinophils, which are stained with acidic stains, basophils which can be stained with basic dyes and neutrophils which take neutral dyes. These are concemed with immunity reactions and defence of body.

Grasshopper. Saltatorial Orthopeterous insect, akin to locust, having biting and cheWing mouthparts; thickened forewings (tegmina) cover membranous hind wings; hemimetabolous.

Grass-snake. Non-poisonous ringed snake, *Natrix natrix* (Ophidia: Reptuia).

Gravigrada. Group of herbivorous xenarthran mammals (Edentata) that are completely hairy and have five upper and four lower prismatic teeth without enamel - sloths etc.

Gray Crescent. Primary organiser. A portion of gastrula inducing formation of ectoderm, endoderm and mesoderm. It becomes dorsal lip of blastopore during process of gastrulation.

Grebe. Fresh-water diving bird, *Podiceps* spp. (Podicipitidre:Aves); flattened head, lobed feet, short wings and no tail - great-crested grebe. P. *cristatus*.

Gland. Modified nephridium like structure functionalin crabs and cray-fishes.

Gregarious. Living In company, Flocks, herds etc. as seen In locust.

Ground Beetles. Dark coloured active. nocturnal insects (Carabidae: Coleoptera), run rapidly on ground and are predaceous on others Insects.

Grouse. Moor-fowl: gallinaceous game-bird of moore-lands. having short curved bill, short legs and feathered feet - *Lagopus mutus*. L. *Scoticus*. L. *tetrix* etc.

Growth. Irreversible Increase In size and dry weight of body cell due to physical and physiological activities.

Grub. Lalva of Coleopterous insects having fleshy wrinkled, curved body with small brownish head and apodous or 3 pairs of thoracle legs but no abdominal legs.

Gryllotalpidae. Brunner, family of male-crickets, substerraneam insects. brownish In colour, short anteunac-eyesuduced I broad spade like toothed forelegs adapted for digging, wings short folded over abdomen. Ovipesiforvestigial cerci long; male produce sound by rubbing wings one over another.

Grysbok. Small South American antelope, body ruddy chestnut coloured with white hair.

Guacharo. Nocturnal, frugivorous oily bird of South America, allied to goatsucker, eaten by local people.

Guan. Large arboreal noisy game-bird (Galliformes : Aves) found in South America.

Guana. Large arboreal lizard (Lacertilia: Reptilia) of South America. GUANACO : wild llama, *Lama guanicoe* (Aruodactiyla: Mammalia) having reddish-brown wool, found in South America.

Guanine. Nitrogen base that combines with deoxyribose sugar and phosphoric acid to form one of the nucleotides making up RNA and DNA. In addition the hydrolysis of GTP (guanosine triphosphate) releases energy to drive some energetically unfavourable reactions (see nucleotides). Chemically it is 2-amino-6-oxypurine.

Gudeon. Small European fresh-water fish, *Gabio gablo* (Pisces), about 12 mm long with only two barbels; bottom dweller; used as fish-bait.

Guenon. Colourful long-tailed monkeys, *Cercopithecus* spp. of Africa -blue-monkey, Diana-monkey green-monkey (vervet). red-tailed- monkey etc.

Guillemot. Diving auk-like sea-bird, *Una* spp. (Aves), swims like duck but on land stands upright like penguin; nests in large colonies on clilts; common in northern seas.

Guinea Fowl. Guinea-hen gallinaceous bird *Numida* spp. (Aves}. of African origin, dark-grey coloured with white-spotted plumage; domesticated in Eurasia.

Guitar-fish. Small guitar-shaped cartilaginous fish (Pisces), 1.5 to 2.0 metres long; head and body broad and flat, tail cylindrical; glides

Gulls. Sea-mew: common name for some sea-birds. *Lams* spp, (Laridc:e : Charadriiformes)that are long-legged. web-footed having brightly coloured beak and peculiar harsh cry - black-headed gull (Lridtbundus), great black-backed *(L.marinus),* lessen black-backed *(L.jilscus).* hen1ng gull *(L.argenttatus)* etc.

Gunylate Cyclase. Enzyme that converts guanosine triphosphate (GTP) anto c-GMP. (cyclic-guanosin3 monophosphate.

Guppy. Small tropical fresh-water fish.. used mainly for aquarium,,; male brightly coloured, 2 to 3 cm long; female drab and about 5 em long; viviparous.

Gurnard. Gurnet: marine fish. *lligla* spp. (Pisces), large wedge-shaped spiny-head and three finger-like walking rays in front of pectoral fin; found in shallow coastal waters.

Gustation. Sense of taste. Chemoreception of ions and molecules in solution by specialised epithelial sensory receptors. Tongue and taste buds present on its surface respond to this sense.

Gustatoreceptors. Organs of taste. Consist mainly of taste buds present over the tongue (see taste buds).

Gymnure. Moon-rat; shrew-like insectivore mammal: haVing thick under fur with dense covering of coarse hair and naked tail.

Gynaecology. Science pertaining to women's diseases; their causes and cure.

Gynecomastia. Development of breast in male. SeenAn mild transient form in many normal boys at puberty. May also be due to a variety of seemingly unrelated conditions including enuchoidism due to primary testicular disease hypothyroidism. hyperthyroidism and cirrhosis of liver.

H

Habitat. Environment in which animal lives. May be aqoratic which means dominated by water. terrestrial means of normal land. subterranian means inside earth and arboreal means on trees.

Haddock. *(Finnan)* Greyish-brown with white belly, edible sea-fish. *Melanogrammus aeglefins* (Pisces) of North Atlantic, that grows upto one metre; characterised by three dorsal fins and dark patch just behind gills.

Haemal. Blood or blood vascular system.

Haematocrit. Percentage to blood volume occupied by erythrocytes. In man normal percentage of haematocrit is 40 to 50 percent.

Haematology. Branch of zoology that deals with blood- its composition, formation, functions and diseases.

Haematuria. Presence of blood in urine.

Haeme. Iron containing red pigment chemically a porphyrin derivative. Part of haemoglobin, myoglobin, methanoglobin and cytochrome-c.

Haemianopsia. Half-blindness. Leisions affecting optic chiasma such as pituitary tumours expanding out sella turica cause destruction of fibres from both nasal hemiretinas and produce this pathological condition.

Haemochromatosis. Disease with large ferritin and haemsiderin deposit and charactertsed by pigmentation of skin pancreatic damage with diabetes. cirrhosis of liver, high incidence hepatic carcinoma and the gonadial atrophy (non functioning of gonads)

Haemocoel. Portion of body cavity functioning as part of circulatory system as seen in insects like cockroach etc. It is filled with haemolymph.

Haemocanin. Bluish respiratory pigment found in the insects' haemolymph. serves to transport oxygen to body tissue.

Haemocytoblast. Primary cells of red bone marrow from which all kinds of blood corpuscles can be derived: irregularly shaped with large deeply staining nucleus and clear cytoplasm. Haemocytoblasts destined to give rise to RBCs are called haemorythroblasts. those giving rise to granulocytic WBCs are called myeloblasts, those giving rise to agranulocytlc WBCs are called nomoblasts and "lynphoblasts, those giving rise to platelets are called megakaroblasts.

Haemoerythrin. Invertebrate respiratory pigment that is a protein but does not contain haeme.

Haemoglobin. Pigment in RBCs of vertebrates, capable of carrying oxygen in the body. In invertebrates it is present dissolved in plasma. It combines with oxygen to form oxyhaemoglobin and when the blood goes to tissues with this compound, it is decomposed and the oxygen liberated is used for oxidative part of respiration.

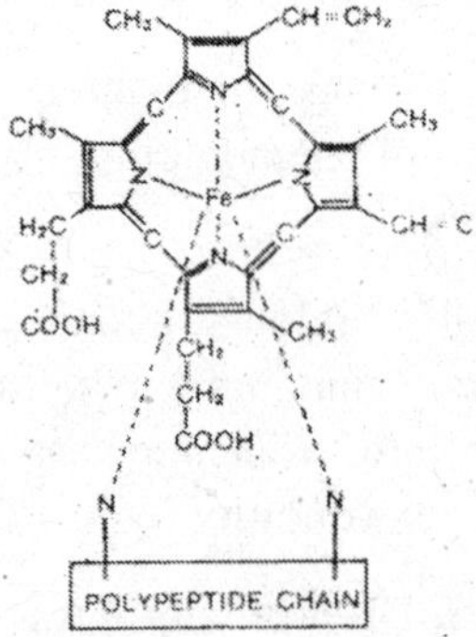

Haemophilia. Abnormal inherited condition in man in which there Is delayed clotting of blood It is sex-linked disease and usually common among males.

Haemopoisis. Process of RBCs formation in vertebrates (occurs in red bone marrow In adult mammals and liver and spleen in amphibia etc). During this process haemocytoblasts differentiate into proerjthroblasts and these cells transform into early and late normoblasts which change to reticulocytes and ultimately give Jise t.o RBCs.

Haemopoitic Tissue. Tissue In which RBCs are formed. Red bone marrow in adult mammals while spleen. kidney and liver are in embryonic stages of mammals and lower vertebrates.

Haemorrhage. Bleeding from some internal. organ/part of the body assodated with defective clotting mechanism in the body (see clotting mechanism). Also due to defective capillary contracitibility.

Haemostasis. Damage in a small vessel which initiates a series of events that lead to the formation of clot, resulting in stoppage of bleeding.

Hagfish. *Myxine glutinosa* (Pisces). slimy smooth, scaleless body, with large jawless mouth surrounded by tentacles; tongue carries many horny teeth; no paired fins; feed on dead or dying fish.

Hair. Slender Hlamentous skin accessory of mammals and of exposed surfaces of some arthropods.

Bake. Gadoid food-fish found in Mediterranean sea; *Merluccius merluccius* (Pisces), brownish-grey dorsally and silvery-white sides and belly. characterised by one short and other very long dorsal fin.

Hallux. First digit of foot. It is absent in rabbit and present in some birds and other mammals like primates. May also be oppasable like thumb *(see pollex).*

Hamadryad. King cobra *Naja bungarus* (Ophidia: Reptilia) of India. Also large Arabian baboon. *Comopithecus hamadryas* (Anthropoidea: Mammalia). worshiped by ancient Egyttans.

Hammer Head. Heron-like bird *Scopus* species (Aves) with large crest on back of its head and broad, fiat beak. ***Shark***: *Sphrya blochi* (SalachU : Pisces) having sides of its head drawn out to form hammerhead. fast swimmer, preys upon small fishes.

Hamster. Group of small, snort-tailed rat-like hibernating rodents *Cricetus cricetus* and allied species with cheok pouches reaching almost the shoulders and it carries food in these: pouches.

Harrier. Hound (Carnivora: Mammalia) used for hunting hares. Kind of long-tailed falcon. *Circus* spp. (Falconiformes : Aves) that harries small animals. ***Eagle***: *Circaetus gallicus* (Accipitridre: Aves) and allied species.

Hawk. Large short-winged hawk *Micrastur* species (Falconi-formes) native to America

Hartebeest. South African antelope. *Alcelephas* spp., large strange looking mammal; long narrow straight face. back sloping sharply from shoulders to hindquarters; both sexes having doubly curved horns; reddish in colour with shoulder height of about 1.7 metres.

Harvestman. Harvest spider or daddy-long-legs is small brownish spider-like arachnid with very long slender eight legs; camivorous, feeds on other small creatures.

Harvest Mouse. Tiny mouse (*Rodenita* : Mammalia) weighing 6 to 7 gm. with soft furry body and long prehensile tail; swings about on grass stems.

Haversian Canals. Canals present in the bone that permits passage of blood vessels and nerves through it. Characteristic structure in the mammalian bone. Haversian canals and Volkmann's canals. The lamellae of bone (osteon) containing osteocytes. surround Haversian canal and the various Haversian canals are interconnected together with the help of Volkmann's canals.

Hawk. Common name for birds of falcon-family (Folciformes: Aves) large. strong. diumal birds of prey. ***Moth*:** Sphinx-moth, large stout insect (Lepidoptera: Hexapoda); caterpillars smooth haVing characteristic dorsal horn on VIII abdominal segment; voracious leaf-defoliators and. destructive to various economic crops and trees.

Heavy-Meromyosin. One of the retions of heavy-meromyosin. It has maximum ATPase activity and helps in cross-linking of the cross-bridges dUring muscle contraction.

Hectodactylus. Male cephalopod mollusc's modified arm used mainly for transfer of sperm to female.

Hedgehog. Nocturnal insectivore mammal. *Hemtechinus collarts,* having pIg-like snout, long ears. prickly coat of spines on dorsal surface and hairs of ventral side: capable on rolling itself into a ball with spines all over; found in Eurasia and AfrIca.

Helicoid. Gastropod shell (Mollusca) looking like flattened spiral or flat coil.

Helicoplacoidea. Extinct class of free-living, pear-shaped echinozoans (Echinodermata) that lived during Cambrian period in western USA

Heliotropism. Response in organism with respect to Sunlight. HELIX: any structure having spiral shape *e.g.*, shell of snaillike *Pila,* DNA helix, cochlea of mammalian ear etc.

Hellbender. Giant American salamandar. Cryptobranchus alleganiensis (Urodela : Amphibia). flattened brown body, small eyes; found in swift streams of USA. China and Japan.

Helminthology. Scientific study of parasitic worms. including_ PlatyhelmInthes. Nemathelminthes etc.

Hemibranch. In fishes the g1lliamella on each side of gill arch. The two such lamellae belonging to one gill arch and the interbranchinal septum in between them constitute a complete gill. Only a single gill lamella is called half gill or hemibranch. The first gill in fishes like *Scoliodon* spp. supports a hemibranch.

Hemichordata. Deuterostomical enterocoelomate animal phylum to which *Balanoglossus* (acorn-worm) belongs.

Hemimetabola. Insects having simple or incomplete metamorphosis - egg. nymph and adult.

Hemione. Common name for several wild asses *Equus* spp. including. dziggetari. kiang. kulan etc., found in various parts of Asia.

Hemipenes. Copulatory organs found in reptiles like, lizards and snakes etc. There is one pair. lodged into a sac-like structure and consists of a pedicel or a stalk bearing a groove along its length. also bealing a glans-like structure at its tip sometimes. Used for transferring sperms into the reproductive tract of female.

Hemiptera (Heteroptera). Order of insects (Hexapoda) that includes true bugs having piercing and sucking mouthparts; forewings hemelytra and hind wings membranous - coreid bugs. painted bugs. tlngids etc.

Hemizygous. Having particular loci present in just a single dose as in case for all sex-lined genes of human male.

Hen. Female of several gallinaceous birds especially cock etc. *Hensen's Node* : anterior end of primitive streak in the embryo of bird.

Heparin. Protein containing molecule capable of preventing blood coagulation. Anticoagulant in the body secreted by Mast cells of areolar tissue (which is found around main blood vessels aprtae etc. and among the visceral organs in the septa and mesenteries). Prevents thrombosis of blood inside the blood vessels during its flow through them. (see areolar tissue).

Herbivorous. Organism feeding chiefly upon plants and vegetable matters - rabbit. elephant etc. In ectosystem these animals are considered to be primary consumers.

Herdarian Gland. Modified sebaceous glands placed behind the eye-balls. Present generally in amphibia. reptiles and birds and rarely in mammals. Their secretion lubricates nictitating membrane.

Heredity. Tendency of offsprings of an organism to resemble their parents. The science dealing with the study of inheritance. It incorporates several laws and principles that are concemed with transfer of traits (characters) from parents to offsprings as for example Mendel's laws, crossing-over, sex-lined inheritance etc.

Heretability. That proportion of phenotypic variation in expression of characteristics which is due entirely to genetic component.

Hermaphrodite. Organism combining characterIstics of both male and female sexes - common in plants and some lower animals like earthworms etc.

Hermaphroditis. Possessing both reproductive organs in one single individudal as seen in earthworm. *Fasciola, Taenia. Hydra* etc., should not be confused with gynandromorphism where part of individual's body exhibits male appearance and part of body as female.

Hermit Crab. Soft-bellied crab (Paguridre : Crustacea) that lives in mollusc's cast-off empty shell to protect its soft shell-less hinder pa

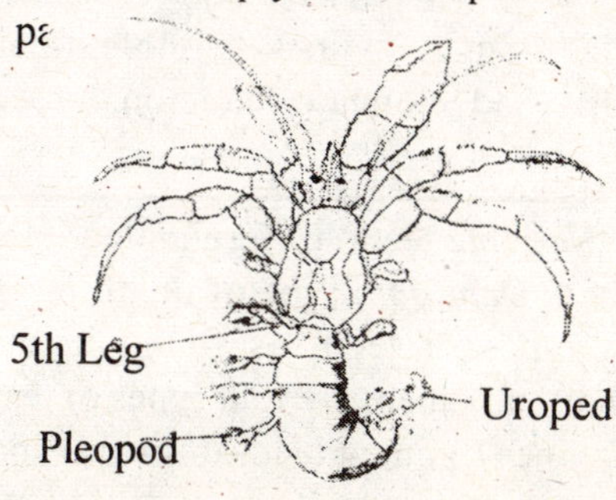

Heron. Large wading-bird Arda cinerea (Ardeidae : Ciconifonnes). long-necked. long-legged with screaming habit - grey heron common in Eurasia. blue heron and green heron are found in North and Central America.

Herpetalogist. Zoologist who has specialised in study of reptiles. HERPETALOGY: branch of zoology dealing with study of reptiles. HERRING: most famous food-fish *Clupea harengus* (Pisces) found in North Atlantic. greyish-green dorsally and silvery beneath; comes to coast in large shoals to spawn in shallow water. *Gull-* arge white gull. Larus *argentatus* (Charadriifonnes) of North America having white wings with black tips.

Hesionid. Family of small polychaete wonns (Errantia : Annelida). *Hesperornithiformes*: order of ancient extinct birds (Odontognathae : Neornithes); that were large but flightless. having shoulder girdle and wings much reduced. legs adapted for swimming; very good swimmers and divers - *Hesperornis* species.

Heterocera. Formerly recognised as suborder of Lepidoptera (Hexapoda) that include all moths (antennre not-clubbed).

Heterocercal Tail. Type of tail which is asymmetrical-internally as well as externally. In such a tail the axis (portion of vertebral column is bent upwards and the two lobes of tail fin (epicaudal/upper lobe and hypocaudal/lower lobe) are of unequal size. Such a tail is found in sharks like *Scoliodon* and bony fishes like sturgeon.

Heterocorallia. Small extinct, monofamiliar order of fossil corals that had elongate skeletons; found in calcareous shales.

Heterodont. Variable teeth. differentiated into incisors. canines and molars. *Dentition*: type of dentition in which there are different types of teeth as found in mammals. In such dentition usually there are four types of teeth-incisors (front-teeth), canines (pointed teeth following incisors). premolars and molars (both grinding type teeth). Teeth in this dentition may be monophyodont (Like molars) or diphyodont (Like other remaining types). The equation representing heterodont teeth in a mammal is called dental formula (see dental formula).

Heterodontoidea. Suborder of sharks (Selachii: Pisces). represented by single living gonus - *Heterodontus.*

Heterogamy. Union of two gametes that differ in size and structure: alternation of differently organised generations - sexual reproduction and parthenogenesis occurring alternatively.

Heteromorphosis. Regeneration of a new part that differs from the part it replaces.

Heteronomous. Condition in which metameres of an animal are not similar: metameres specialised in various parts of the body.

Heteroptera. (*Hemiptera*): order of true bugs (Hexapoda). insects of variable size. mouthparts modified to form beak or rostrum: forewings partly thickened. partly membranous: metamorphosis gradual - assassin bugs, lace bugs. plant bugs, stink bugs etc.

Heterosexual. Person sexually attracted to opposite sex of same species.

Hetrosexual Behaviour. Sexual drive of an individual towards individual of opposite sex. Opposite to homosexual behaviour.

Heterosexuality. Performance of sexual act by an individual of opposite sex of the same species. Opposite to homosexuality.

Heterothermix. Animals who do not maintain constant body temperature but exhibit endothermic regulations during their activity period - protothria and metatheria among mammals.

Heterotroph. Organism having different ways 'of nutrition except autotrophic one; such ways include holozoic. saprozoic and parasitic methods of nutrition which are seen commonly among most of free living and parasitic animals.

Heterozygous. Individual having two members of allelic pair of genes which are dissimilar for example tall hybrids having genetic constitution Tt or tT, thus produce two kinds of gametes one with factor-T and other with factor-to Opposite to homozygous.

Hexacanth Embryo. Embryo of some tape-worms like *Taenia solium* bearing six hooks and is surrounded by protective layers. Found on faecal matter of primary host (man). It is infectious to secondary host (pig) and mode of infection is oral. In pig, it settles in voluntary muscles by changing into cysticercus larva.

Hexokinase. Enzyme that catalyses phosphorylation of hexose sugars like glucose.

Hierarchy. System of ranking living organisms; classification indicating the categorical levels of various taxa. *Hind*: posterior (region) ; female of stag. *Gut*: in arthropods. posterior portion of digestive tract; lined with cuticle. Also in the invertebrate's embryo, rudiment of posterior portion of digestive tract.

Hinny. Hybird quadruped mammal resembling horse but smaller in size; offspring of horse and she-ass Oennet).

Hippocampal Commissure. Median connection joining two hippocampal lobes in mammals often it is called psalterium of lyra.

Hippopotamus. River-horse, large, aquatic, artiodactlyl ungulate. herbivorous mammal *Hippopotamus amphibious*, having heavy thick-skinned hairless body with large head and muzzle and short legs; native of Africa.

Hirschprung's Disease. Diseasc affecting young children; there is degeneration of cells of Auerbach's plexus in region of pelvirectal junction, often great length of colon may be involved; affected portion fails to transmit waves and constitute a junctional block; colon above block becomes greatly distended. There is a marked constipation and great abdominal distention followed by attacks of in~estlnal obstruction.

Hirudinea. Class of parasitic or predatory annelids; body elongated oval and flattened dorsoventrally. 34 somites or body segments and terminal suckers for attachment to host and locomotion; hermaphrodite -leeches.

Histogenesis. Origin, development and differentiation of tissues of an organism.

Hitotoxic Hypoxia. Condition of oxygen deficiency at tissue level in which amount of oxygen delivered is adequate but due to action of toxic agent the tissue cells cannot make use of oxygen supplied to them.

Hoatzin. Tree-climbing, occipital crested bird found in forests of South America. reptilian ancestory is evident from presence of claws~on wing-tips of young ones.

Hodgkin Cycle. Regenerative or positive-feed back loop responsible for uptake of action potential. Depolarisation causes an increase in Na+ permeability permiting an increased influx of Na+ which further depolarises the membrane.

Hog. Common name for various species of pigs and castrated boars (Suidae : Mammalia).

Holandric Inheritance. Sex-linked inheritance where 'Y' chromosome contains genes in question. *e.g.,* hairy ears in males have been defined to be inherited holandrically in man.

Holocrine Gland. Gland in which the cells when filled to capacity with products die and are discharged from epithelial wall and new cells replace the discarded ones *e.g;* oil glands or sebaceous glands in mammalian skin.

Holozoic. Nutrition found in most animals. It involves ingestion of food, deglutition, digestion, absorption, assimilation and egestion of undigested material.

Homoestasis. Living organisms create and maintain their steady state at the cost of their environment and always have a tendency to remain in optimally tuned condition. This tendency has been defined as homeostasis. Term coined by Claude Bernared as 'rnileu interior' which means internal environment.

Homocercal Tail. Type of tail which is externally symmetrical but asymmetrical internally. Such a tail is found in most bony fishes - *Labeo* species.

Homodont. Dentition having all teeth of same type as in Amphitia, Reptilians etc. cf. heterodont.

Homodynamic. Developing through continuous successive generations without diapause, as in insects.

Homoeclous. Having only one host during all stages of an organism's life-cycle.

Homogametic (*Monogametic*). Producing only one type of gamete with regard to sex chromosomes; in most animal species (including man) females are homogametic.

Homogenesis. Mode of reproduction in which offspring is like parent and passess through same cycle of existence.

Homologous. Chromosomes carrying genes for same traits, one is paternal being transferred from father and the other being transferred from mother. These chromosomes come to unite or pair during synapsis of meiosis.

Homology. Basic similarity; structural likeness of an organ/part of one animal with, comparable unit in other resulting from common ancestry. These organs may/may not have same function *e.g.*, wings of birds have homology with paired fins of fishes although the wings help in flight and the fins help to position the body iJ? the direction of locomotion.

Homoneura. Suborder of Lepidoptera (Hexapoda); insects having mandibulate mouthparts; fore- and hind wings similar in shape and venation.

Homoptera. Order of sucking insects (Hexapoda); beak, if present, arises from back of underside of head; wings, when present, of uniform texture; metamorphosis (except in male scale insects) incomplete - aphids, scale insects, whiteflies etc.

Homo Sapiens. Modem man (Hominidre: Primates) - omnivorous, terrestrial, erect, biped.

Homosexual Behaviour. Sexual drive of an animal/individual towards another individual of same sex (of same species). Opposite of heterosexual behaviour.

Homosexuality. Performance of sexual act by an adult with a member of same sex within same species. Opposite to heterosexuality.

Homothermous. Keeping body temperature constant-all warm-blooded animals.

Homotype. Specimen compared with type by a person other than original author and determined as conspecific with the type.

Homozygous. Individual in which the members of a given pair of genes are alike/similar; *e.g.,* hybirds with genetic constitution IT or tt. Opposite to heterozygous.

Bear. KinkaJau : carnivore arboreal mammal.

Bees. Social insects (Apidre : Hymenoptera), live in colonies (bee-hives), yield honey and wax; help in pollination.

Buzzard. Falcon, *Pemts aptvora* (Falconiformes : Aves), bird of prey that feeds on bees and wasps. *Eaters*: any of group of song-birds (Aves) found in Australasia; having slender down-curved beak for sucking; extremely fond of nectar, birds are attracted to various flowering shrubs but insects (on wing) make up greater part of their diet; the Flocks are conspicuous by their noisy chattering - Holland-honey-eater *Phylidonyrts novahollandiae.* spiny-cheeked honey-eater *Acanthagenys ruJogularts.* white-eared honey-eater *Meliphaga leucotis.* yellow-faced honey-eater M *.chrysops,* yellow-plumed honey-eater *M.omata* etc. *Mouse*: mouse-like, long-tailed marsupial mammal *Tarsipes* species of South-west Australia, characterised by having pointed tubular snout and long bristly tongue; feeds on insects and nectar.

Hoof. Specialised structure found in ungulates like, cow, sheep, goat etc. which is adapted for walking on hard surface. It consists of unguis or sole plate which is thick and curved around the end of digits and encloses subunguis or sole.

Hooklet. Hooked structures that hold barbs of a feather of a bird and help in formation of vane in contour feather. These are absent in down feathers and flloplumes etc.

Hook-Worm. Common name for nematode parasite of man *-Ancylostoma duodenale.*

Hoopoe (*Hud-Hud*): familiar, graceful insectivorous bird, *Upapaepops* (Piciformes : Aves), fawn-coloured with black and white markings on back, Wings and tail; conspicuous variegated plumage, large erectile many feathered head-crest and slender curved beak, helpful in pulling out insects from holes in tree trunks etc.

Horizontal Cells. Nerve cells whose fibres extend horizontally ln outer plexiform layer of retina.

Horn. Cornified -cap-like outgrowth of skin on the head region of animals like cow, sheep, goats etc. These are of following *types*: *Prong horn*: permanent projection of frontal bone. covered with horny sheath. Shed annually. Found in antelope. *Antlers*: branched horn formed by branching outgrowth of frontal bone which is covered, during growth by hairy skin which is shed annually. Found in deer family. *Giraffe horn* : it is just like antler but the velvet (soft skin covering the horn) is never shed. Short and unbranched. Found in giraffe. *True horn* : formed by the growth of skin of head region and its subsequent cornification. These are found in cattle like. cows. sheep. buffaloes etc. *Fibre horn* : this is made of keratinised cells of epidermis and consists of matted keratin fibres bound together but fibres are not true; also called hair-homs and are found in rhinoceros.

Hornbill. Medium-sized bird *Tockus birosiris* (Bucerotidre. Piciformes) of Asia and Africa. characaterised by extra large beak with brightly coloured hom-like excrescence.

Hornet. Common name for number of large wasps (Vespidre: Hymenoptera).

Host. Organism providing food. shelter or other benefits to another organism without any gains. - *Cells*: epitheliomuscular cells of cnidarians which lodge cnicoblasts.

Housefly. Small blackish flies. *Musca domestica* (Muscidce: Diptera), reddish-brown big eyes. lapping mouthparts and transparent Wings; vector for transmitting various human diseases.

Howler. Largest black and brown monkey. *Alouatta* spp. (Platyrrhina: Primates), fiat-nosed with long prehensile tail; its howling cry can be heard several kilometres away.

Hyaruzi. Offspring of male llama *(Lama glama* ; Camelidce) and female alpaca *(L.pacos)*.

Human Being. Modem man Homo *sapiens* (Catarrhina: Primates). *Genetics*: scientific study of heredity and variation in human beings.

Humming Birds. Smallest, fast-flying birds. *Trochus* spp. (Trochilidce: Aves) having colourful plumage. narrow beak. tubular fringed tongue. short legs and weak feet.

Humoral. Body fluids such as hormones or other chemical agents. HUMPBACKED WHALE: rare, medium-sized toothless whale (Cetacea: Mammalia), about 16 metres long with unusually long flippers; found in coastal waters. tropical in Winter and polar in Summer.

Humoral. Body fluids such as hormones or other chemical agents. HUMUS: organic part of the soil.

Hunting Ton's Disease. Nervous disorder generally manifested well after puberty and characterised by involuntary movements of face and limbs along with a gradual loss of mental faculties.

Hyaena. African wild dog. *Hyaena hyaena.* carrion-feeding carnivore mammal with thick neck. coarse mane and sloping body.

Hyaluronidase. Enzyme produced by sperm that helps the sperm to break through the egg membrane and the same, fertilise by . making its way into the ovum.

Hybrid. Indidividual resulting from union of egg and sperm that differ in one or more genes; offspring of two different species-liger, tigon; hinny, mule etc.

Hybrid Vigour. Exhibition of more vigorous growth. greater yield, increased disease resistance by a hybrid than either parent. The effect is thought to be due to accumulation of dominant alleles each having additive effects and masking effects of deleterious recessive alleles.

Hydra. Soft-bodied coelenterates. 3 to 10 mm long; double-walled (diploblastic) body with mouth surrounded by tentacles and having both male and female reproductive organs (hermaphrodite) .

Hydrated State. Surrounded by water molecules usually occurs because of polarity.

Hydrocaulus. Main stalk-like stem of a hydroid colony.

Hydrocortisone. Glucocorticoid that affects carbohydrate metabolism,yrotein metabolism. lipid metabolism, electrolyte and water balance, relieves muscular weakness. corrects eosinophilia. anaemia, lymphocytosis and neutropenia etc., overcoming myocardial weakness. controls nervous activity, pepsin production. metabolism of bone. infection. inflamation and trauma etc.

Hydroid. Pertaining to hydrozoa; refers to polyp of hydrozoan. as distingUished from medusoid form.

Hydrophilic. Phospholipid molecules in cell membrane contain phosphate portion which is positively charged and is quite soluable in water. In the membrane these ends are exposed to aqueous environment that baths exterior of cells and the aqueous cytoplasm.

Hydrophobia. Morbid fear of water : inability to swallow water. *Hydrophobic.* Tail of phospholipid molecule, of cell membrane that are quite insoluable (non-polar). The hydrophobic ends meet in water-poor interior of membrane.

Hydroxy Tryptamine. Substance formed by hydroxylation of essential amino-acid tryptophan which forms seratonin by decarboxylation.

Hydrozoan. Small colonial creatures belonging to class Hydrozoa (Coelenterata~ showing well marked alternation of generations.

Hymen. Membrane that partially closes virginal vagina. *Hymenoptera*: third largest order of insects (*Oligoneoptera: Hexapoda*); mouthparts often modified for lapping or sucking.; wings membranous with few or no cross-veins ovipositor well developed often modified for piercing. sawing or stinging; metamorphosis complete-ants. bees and wasps.

Hyoid. Group of bones and cartilages at or near the base of tongue.

Hyostylic Suspension. Jaw suspension found among sharks where upper jaw has no direct attachment with cranium but is suspended by ligaments with brain box through hypomandibular which is movably articulated to otic region of cranium.

Hypapophysis. Hypocentrum in some vertebrae of reptiles. birds and animals linger on the ventral side of centrum as a bony nodule which is called hypapophysis.

Hyper. Excessive. more than normal.

Hyperaldosteronism. Aldosterone secretion enhancement in the individual primary when some aldosterone secreting turmours of adrenals' cortex produce excess of mineral-ocorticoids and secondary when it is caused due to nephrosis. congestive heart failure etc. (due to elevated levels of renin and angiotensin-II and may cause increased aldosterone secretion).

Hyperalgesia. Pathological condition in which the sensitivity of the pain receptors is altered. Two types of alterations are seen -primary when pain in the flare (see flare) is noticed and secondary when another aberration of sensation is marked following an injury.

Hyperdactyly. Condition when an animal has large number of digits: found in some aquatic dinosaurs.

Hyperglycemia. Increase in blood sugar level from normal; caused by disturbed insulin production in the body. There is reduced ability to metabolise or convert them into fats and loss of proteins which are broken down for energy in place of glucose. The disease can be cured by administration of insulin; which -induces. glycogenesis so that excess of blood sugar is controlled.

Hyperkalemia. Increase in K+ions of the blood. It is very dangerous and potentially lethal condition because of its effect on heart. It may cause paralysis of atria and prolongation of QRS complexes. Altered repolarisation causes appearance of tall peaked T waves.

Hypermetropia. Inability of eye to see near objects clearly.

HYPEROPIA: near-sightednes, condition where eye-ball is relatively long, leading to the eye's point of focus lying in front of eye instead of retina.

Hyperparasite. Parasite that parasitises another parasite. *Hyperparasitism*: condition where the parasites are parasitised by other parasites; may be secondary. tertiary etc.

Hyperphalangy. Condition when an animal has large number of phalanges as seen in aquatic dinosaurs.

Hyperplasia. Excessive multiplication of cells resulting in outgrowth of that particular part of body.

Hyperpnea. Increased lung ventilation in an animal.

Hypertension. Systolic blood pressure in normal healthy person does not exceed 145-150 mm of mercury. A value above this in otherwise normal and healthy man is hypertension; caused generally during vigorous exercise, emotional disturbance. smoking, over obesity and due to inherited traits. The common symptoms of this disease are headache, throbbing of head, fatigue, dizziness and palpitation. In extreme condition it may cause cerebral haemorrhage giving rise to apoplectic stroke.

Hypertonic. Higher osmotic concentration. In terms of living cell the concentration of water molecules is greater inside the cell than on the outside, after outside, therefore more water molecules pass out of the cell than into it which causes shrinking of the cells.

Hypertrophy. Abnormal increase in size of an organ or the other part of the body.

Hyperventilation. Abnonnally prolongecl. rapid. deep breathing resulting in dizziness or even fainting.

Hypervitaminosis. Disease caused by an excessive intake of vitamins. For example hypervitaminosis of vitamin K causes gastrogenital disturbances and anaemia. Hypervitaminosis of vitamin D causes weight loss and calcification in many soft tissues.

Hypnotoxin. Poison present in nematocysts of cnidarians like *Hydra;* used to paralyse the prey during ingestion period.

Hypocentrum. In embolomerous vertebra the anterior centrum is believed to have been derived from basiventrals. This is called hypocentrum.

Hypocercal Fin. Caudal fin in which notochord or vertebral column extends into ventral portion of tail. The ventral portion extends farther than the dorsal portion of tail.

Hypoconvertinemia. Due to congential cause there is deficiency of blood clotting factor-VII (proconvertin factor) in some individuals as a result of which there is less activation of factor -X (which is responsible for conversion of prothrombin to thrombin) and have low capacity for coagulation.

Hypodactyly. Condition in birds when number of digits is less (three toes in most of the birds and only two in ostrich).

Hypodermis. A layer situated below dennis of skin also called sucutaneous tissue. Composed of connective tissue fibres and cells. In mammals it consists of adipose tissue and fonns a layer called pannaculus adiposus. In aquatic mammals like whales, it is in fonn of a very thick layer of fat called blubber.

Hypoglossal. Twelfth cranial nerve found only In mammals and derived from several posterior ventral spinal roots and supplies the muscles of tongue and below it. Some fibres from the nelve together with fibres of certain spinal nelves form a plexus in the neck which is called ansa hypoglossi from which nelves go to the muscles that draw larynx backwards.

Hypoglycemia. Fall in blood glucose level of body. The brain cannot function and there is often seen 'hypolycemic fit'. Under this condition glucagon is secreted by alpha-cells. of islets of Langerhans which stimulates glycogenolysis that brings down conversion glycogen to glucose so that normal blood glucose level is restored. Patients under hypoglycemic fit require immediate intravenous injection of glucose.

Hypogonadism. Low secretion of gonadotropins due to hypothalmic disease, causes underfunctioning of gonads leading to improper body development, delayed puberty, improper development of secondary sexual characters, improper development of reproductive organs and disturbed reproductive cycles and disorders.

Hypomere. Mesodermal pouch during embryonic development differentiates into two parts; the ventral part of this is called hypomere which forms coeiom in the embryo (dorsal part of the pouch form myotome\ and dermis.)

Hypophagia. Condition in which the animal reduces amount of ingested food material. Caused by leisons in the area nearby the centre in hypothalamus. cr. hyperphagia.

Hypopleura. Sclerite above and in front of hind coxa in true flies (Diptera: Hexapoda).

Hypopyle. Posterior opening of claspers attached with the cloaca in sharks. This opening selves as an exit through which sperms leave the claspers and pass into the cloaca of female.

Hyporachis. After shaft: in contour feather, the small feathers with weak vane and coming out at the junction between calamus and the rachis.

Hypostome. Projection surrounding oral aperture (mouth) of many coelenterate polyps. Raised area on posterior oral margin of crustaceans.

Hypotarsus. Process on tarsometatarsal bone in birds (Aves). *Hypotension*: systolic blood pressure under normal condition does not exceed 145-150 mm of mercury. A fall in systolic blood pressure below normal range is called hypotension. Blood pressure below 100 mm of mercury is indicative of feeble circulation and below 80 mm of mercury means impending death. The low blood pressure is caused due to shocks. severe haemorrhage. burning, hypothyroidism and disturbances of adrenal gland. Tried and worn-out attitude. run down condition and impaired metabolism are the symptoms seen very commonly.

Hypothalamico Hypophysial Relay System. System formed by portal veins linking the capillaries of hypothalamic median with those of adenohypophysis. These veins transport hypothalamic neurosccretions directly to adenohypophysis.

Hypothalamus. Part of diencephalon that forms the floor of median ventricle of brain. It includes optic chiasma. mamillary body tuber cinarium and infundibulum. It is the seat of all visceral activities and secretes a number of substances that induce secretion of a number of pitutary hormones. It also controls appetite, regulates body temperature. controls emotions. Maintains constancy of internal environment (homeostasis), prepares body to face stressful situations. and regulates a number of visceral functions.

Hypotremata. Group of cartilagenous fishes that have ventral gill slits. These are commonly called rays, examples are electric ray, sting ray etc.

Hypoxica. Probiem of breathing in normal individuals at high altitudes and a complication of pneumonia and number of common diseases like paralysis of respiratory muscles. skeletal deformities. decreased elastic recolling etc.

I

Ice-fish. Certain Antarctic fishes, *Chaenocephalus* species (Chaenichthyidre : Pisces) having colourless blood, lack erythrocytes; live in oxygen-rich waters and perform gaseous exchange by diffusion.

Ichneumon. Iberian mongoose, *Herpestes ichneumon* and allied Species (Carnivora : Mammalia) of North Africa; Egyptian species is known to destroy eggs of crocodiles. *Fly:* Solltary wasp belonging to family Ichneumonidre (Hymenoptera: Hexapoda), parasitise other insects specially lepidopterous and coleopterous.

Ichthyologist. Person having sound knowledge oflchthyology. *Ichthyology*: branch of zoology that deals with study of fishes. *Ichthyophagous*: fish-eater; one who lives only on fish diet. *Ichthyopterygium*: lobed fins of extInct fishes like *Sawupterus* and *Eusthenopteron* which were paddle-like. They are supposed to be the type of stnictures from which the tetrapod limbs are likely evolved.

Ichthyornis: Fossil bird of Cretaceous period that had vertebrae like those of fishes.

Ichthyornithiformes. Order of Odontognathae (Neonithes : Aves). ancient fossil birds that lived during upper Getaceous period. had all skeletal characteristics of modem birds and were strong fliers - *Ichthyomts* species

Ichthyosauria: Extinct order (Ichthyopterygia) that included gigantic Mesozoic era fossil dolphin-like marine reptiles.

Icsh. (*Interstitial Cell Stimulating Hormone*) : Hormone from anterior pituitary lobe which stimulates ovulation and lute l nisation of ovarian follicles in female and testesterone secretion in the male individuals.

Icterus. Pathological condition in which skin. scleres. mucous membranes etc. turn yellow OWing to accumulation of free or conjugated bile pigment bilirubin. The condition is detectable when total plasma bilirubin is greater than 2 mg/dl. (see jaundice).

IDE. Freshwater fish. *Leuciscus idus* (Teleostei: Pisces) found in North Europe.

Identical Twins. One egg twins or Monozygotic twins: two individuals developed from a single fertuised egg and therefore having identical set of genes.

Ideotype. Specimen identified as belonging to specific taxon but collected from other than type locality.

Idiopathic Hypertrophy. Hyperfunctioning of any part of body without any distinct pathological reason.

Idiosome. Sex chromosome. IDIOTHERMpUS : warm-blooded (animals).

Idioventricular Rhythm. Beating of ventricles at a slow rate Independent of auricles when conduction from auricles to ventricles is interrupted completely or third degree of heart block is said to be present.

IDL (*Intermediate Density Liporoteins*): substances formed during fat metabolism in the liver by conversion of very low density lipoproteins by loss of some triglycerldes and protein in the complexes.

IGM : IgM immunoglobulin found throughout vertebrates, of much higher molecular weight than IgG.

IGuana. Group of large thick-tongued arboreal lizards. *lyuana* spp. (Iguanidre: Lacertilia) of tropical America, grows upto 5 metres in length (including tail); some species are eaten as food.

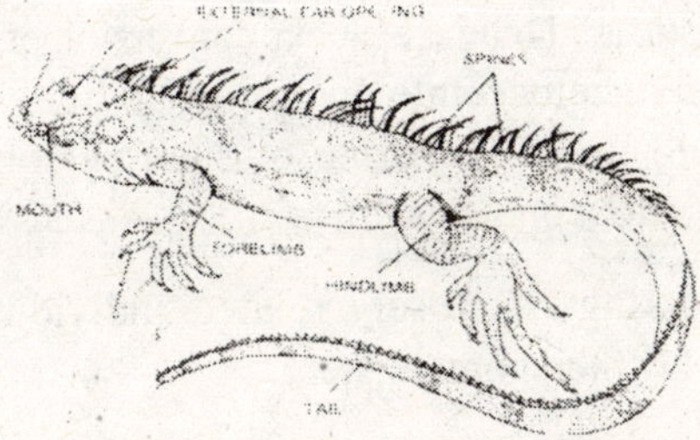

Iguanid. Any member uflguanidre Oacertilla: Reptilia).

Iguanodon. Huge fossil, terrestrial. herbivorous, biped dinosaur (Reptilia) of Jurassic and Cretaceous periods; had teeth and bones like those of Iguana.

ILA (*Insulin Like Activity*). In plasma insulin like activity has been bioassayed by comparing the effects of plasma samples with those of known amount of insulin on a variety of physiological parameters. The ILA values in plasma are generally higher than the values determined by assays, and a significant ILA persists even after removal of pancreas. Normal concentration of insulin measured by immunoassay in peripheral venous plasma of fasting normal humans is 0.70 *J.L* u/ml. The amount of insulin

Implantation. Burrowing of blastodisc (blastocyst) into the endometrium of uterus in case of eutherian mammals. It is due to erosion of endometrium by syncytiotrophoblast of embryo and occurs on the dorsal side of wall of uterus. A placenta then develops and trophoblast remains associated with it.

Impulse. Physiological disturbance created by stimulus transmitted normally along the axon to its termination.

Inborn Metabolic Disorders. Heritable diseases due to some malfunctioning in a gene coding for some specific enzyme involved in a particular step during metabolic pathway leading to physiological disorder. About more than 200 such diseases are known *e.g.*, alkaptoneuria in which defective functioning of homogenistic acid oxidase blocks conversion of homogenistic acid into melylacetoacetate due to which homogenistic acid is passed into urine and urine of such person turns black on corning into contact with air.

Inbreeding. Crossing of closely related individuals *e.g.* first cousins or brothers and sisters.

Incirrata. Suborder of Octopoda (Cephalopoda: Molusca).

Incomplete Dominance. Partial dominance. Term used in genetics for cases where heterozygotes are distinguishable from homozygotes.

Incubation. Act or process of hatching eggs - natural or artificial. ***Incubator***: brooding hen: apparatus for hatching eggs by artificial heat; also for rearing prematurely born infants.

Incus. Mammalian ear- ossicle: anvil.

Independent Events. Two or more alternatives whose separate occurences do not influence one another in any way.

Indigenous. (*organisms*) native to particular area or region and not introduced from other places.

Indri. Rare wooly lemur (Lemuriformes : Mammalia)of Malagasy; about 75 cm long with tailless than 2 cm; fur black and white; active during days. it moves by leaping; (eeds on leaves.

Inosine Triphosphate. One of the phosphorylated compounds having pyrimidine or purine other than adenine. It contains inosine.

Insect. Bilaterally symmetrical arthropod animal (Hexapoda); body comprising of head. thorax and abdomen: adult having three pairs of jointed legs (without exception) and usually two pairs of wings attached to thorax - bugs. flies. moths etc.

Insect (Hexapoda). A class of Arthropoda. INSECTICIDE: toxin that can kill or combat insects.

Insertion. Place of attachment of a muscle to a movable part in contrast to the origin which is the point of attachment to a relat;ively immovable part.

Inspiration. Intake of air into lungs. Also the impetus for creative work.

Instinct. Specific set of unlearned behaviours and responses; inherited response. often complex. invoked by a particular stimulus and leading to a particular response.

Insulin. Hormone secreted by β-cells of islets of Langerhans controlling blood sugar. It is responsible for glycogenesis of excess of sugars and the insufficiency causes diabetes mellitus.

Insulinoma. Insulin secreting tumour which causes chronic elevation of insulin level. Affects glucostasis in the body.

Integration (Neural). Synthesis of an output based on the sum of inputs to a neuron or neural network.

Integument. Outer covering specially the skin of a vertebrate along with its derivatives. Consists of epidermis which is derived from ecotoderm and dermis which is mesodermal in origin. Besides it includes various structures like hair, feather, scales, plates etc. present on the skin.

Intercalated Discs. Cardiac muscles' fibres are divided into shorter cylindrical units by transverse partition which are called intercalated discs. Electron microscopy shows that there is no continuity of protoplasm in this region but they are formed by interdigitation of cell processes from either sides. They represent relay stations for transmission of waves of contraction from one fibre unit to the other.

Intercellular. Substance present inbetween cells. *Intercolumnar Tubercle*: one of the areas outside the brain stained when acidic dye is injected to animal. There are fenestrated capillaries and many appear to contain relatively large quantities of hypothalamic hormone. Functiohs as chemoreceptor.

Intercostal Muscles. Muscles that help to operate the ribs in mammals and enable lowering of sternum during inspiration and expiration.

Intermediary Metabolism. Interrelationship between carbohydrate, protein and fat metabolisms. This enables availability of required metabolite in the body immediately irrespective its dietary intake by the individual.

Internal Secretion. Hormone, neurosecretion etc. from some ductless gland or tissue.

Interorbital Septum. Bony septum between two orbits found only in birds and mammals. Such a skull in which it is present is called tropibasic skull in contrast to platybasic skull in which it is absent.

Interoreceptors. Receptors/sense organs found in the body which detect temperature, pain, hunger, thirst, fatigue and muscular position. Opposite to exteroreceptors.

Interphase Cell. Cell that is not undergong mitoisis: as metabolising cell; stage between two mitotic divisions.

Inter-Renal Bodies. Endocrine structure of fishes lying between kidneys as a pair of glands or single gland.

Intersex. Abnormal individual; intermediate between two sexesin characteristics; not necessarily hermaphrodite.

Intervertebral Discs. Fibrocartilagenous disc-like structures between two centra of vertebrae in mammals. The central portion of sic is called nucleus pulposus an represents remanants notochordand disc probably represents hypocentrum.

Invertebrate. Animal lacking a backone or vertebral column – annelids, arthropods, molluscs etc.

Investing Bones. Bones developed in connective tissue also called membrance bones e.g., frontoparietal, parasphenoid, nasals, vomar, septomaxillar, palatine, pterygold, maxillae, premexillae, qudratojugal, angulosplenial, dentary, mentomeckalian in the skull of frong.

Involuntary Muscles. Elongated muscles pointed at both ends (200 μ long and 6 μ wide) and a single nucleus inthe middle. Saroplasm contains myofibrils. Function independently and never foundin bundles. These are present in urinary bladder, muscle layer of intestine etc.

Iodine. Essential element to body and is involved in biosynthesis of throxine. Its deficiency in diet causes goitre.

Iodopsi. One of the cone pigment mostly sensitive to red light in human beings, made up of retinine and photopsin.

Iodotyrosine Dehydrogenase. Microsomal enzyme that deiodinates the iodinated tyrosines. The iodine liberated this way is reutilised.

Ionic Bond. Bond between two or more atoms formed because of attraction of oppositely charged atoms.

I.Q. (intelligence Quotient). Ratio of mental age and chronologicalage of a person multiplied by 100. This helps to decide the intelligence of persons. I.Q. of some types of human beings:

140-or more	–	Gentus
120 - 139	–	Most superior
110 - 119	–	Superior
90 - 109	–	Average
80 - 89	–	Ordinary
70 - 79	–	Dull
50 - 69	–	Moron

Isoantigen. Antigen found in only some individuals of a particular species.

Isoenzymes. Different molecular forms of proteins with some enzymatic activity.

Isogamy. Sexual fission involving gametes of opposite sexes but structurally alike.

Isoptera. Order of social, polymorphic insects (Hexapoda) that live in colonies (termitaria); mandibulate mouthparts, eat almost everything containing cellulose, starch or sugar; several castes - queen, kinds, soldiers and workers - termittes or white ants.

Isorhizas. Anematocysts of Hydra spp., used of or adhering the body with substratum or for offence. Two types - holotrichous isorhizas which have oval body and long theat with rows of spines and atrichous isorhizas which are slightlysmall but the thread has no spines.

Isotope. Any of two or more forms of an element differring from one another in atomic weight but not in chemical properties. Variation in neutron number. May be radioactive.

Ithomiin. Subfamily of Nymphalidae (*Lepidoptera*: Hexapoda), weak-flying butterflies having broad glossy-wings with transparent areas that are covered withshort hairs instead of scales.

Itonididae (Cecidomyiidae). Family of Nematocera (Orthorrhapha: Ditera); antennae longer than head and thorax; arista wanting; includes gall-midges, crop posts that cause galls on the affected crops.

J

Jacamar. Insectivorous bird (Aves) found in Central and South America; it has long bill. iridescent plumage. reversible fourth toe and long tail.

Jacana. Water pheasant or lotus bird; tropical wading bird(Jacanidre: Aves) having large feet (toes) that enables· the bird to run about on floating plants (lilies etc.).

Jacinth. Beautiful. slaty-blue, fancy pigeon (Columbiformes: Aves).

Jackal. Wild gregarious wolf-like. camivorous mammal. *Canis aureus* found in Asia and Africa; feeds on carrion and small animals.

Jackanapes. Small tamed monkey (Mammalia).

Jackass. Male ass or donkey.

Jackdaw. Small crow, Corms *monedula* (Passeriformes: Aves) found hiding in old and dilapidated buildings.

Jack-Rabbit. Large praire-hare (Lagomorpha: Mammalia) having very long ears and legs.

Jacobin. Species of pigeon (Columbiformes: Aves) having reversed feathers on back of neck that look like hood-shaped covering.

Jacobson's Organ. Pair of sacs with strongly pigmented walls and supported by cartilagenous structures; contain sr.lelling apparatus and smells food when it is in oral cavity; connected to oral cavity with nasopaltine duct.

Jaguar. Large (about two metres long), camivorous quadruped mammal. *Panthra onca* found in forests of Central Africa.

Jaws. Structures normally used for holding or chewing the food materials. Found in gnothastomates like fishes. amphibia, reptiles. birds and mammals.

Jaw Suspension. Splanchnocranium in ganthostomates (vertebrates with jaws) serves to suspend the lower jaw with cranium and forms jaw suspension. The different types of jaw suspension are : *Amphistyllc* : upper jaw Is attached directly to otic region and hyomandibular is movably attached to otic region and slings upper jaw with cranium. Found in primitive sharks. *Hyostylic* : upper jaw has no' direct attachment with cranium but suspended by means of ligaments to it through hyomandibular which is movably attached to otic region. Found in sharks. *Amphidiastylic*: upper Jaw is seen attached (by ligaments) with cranium and the hyoid arch is not in the service of Jaws. Found in acanthdi fishes. *Autostylic* : upper Jaw is fused with the cranium and lower Jaw is articulated with quadrate. Found in amphibia. reptiles, birds etc. Two subs types of this can be differentiated - monimostylic when quadrate is immovable as found in most lizards and streptostylic when quadrate is movable as found in snakes.

Jay. Noisy chattering bird *Ganulus glandarius* that has fawn body plumage with white patches. blue and black wings and black and white erectile crest.

Jelly-Fish. Common name for a scyphozoan coelenterate *Aurelia* spp.

Jerboa. Small noctumal jumping rodent. *Dipus* species found in deserts of North Mrica having short forelegs and long tufted tail. Australian marsupial, *Antechinomys* species.

John Dory. Flat, circular (about 60 em. diameter), golden-yellow, marine fish (Pisces) having spiny dorsal fin and short tail.

Jugal. One of the investing bones of skull forming part of upper jaw in frog; also forms upper jaw with premaxillae. maxillae in mammals; forms slender bar with maxilla and quadratic jugal joining quadrate posteriorly in reptiles.

Jugular. Pertaining to neck or throat: (of fish) having ventral fm in front of pectoral - *Blenntus* spp. *Artery*: artery pertaining to throat. *Pulse*: pulse felt in large jagular vein (in neck region); can be recorded by using suitable tambour applied to neck skin; it has three waves - a *wave* is-due to atrial systole, *c-wave* is transmitted manifestation of rise in atrial forming pressure produced by the bulging of tric'.1spid valve into atria isometric ventricular contraction and *v-wave* mirrors indicate rise in atrial pressure before tricuspid valve opens during diastole. Jagular pulse has clinical importance. *Vein*: vein pertaining to throat.

Juliformia. Millipedes (Diplopoda: Myrtapoda) with 45 or more cylindrical segments each bearing a pair of legs; feed on decaying vegetable matter.

Jurassic. Geological period found 180 million years ago and existed for 45 m1llion years; dominated by dinosaurs: birds also originated during last phase as evidenced by *Archaeopteryx*.

K

Kallidin. One of the kinins formed in plasma. Its name is lasylbradykinin which is a decapeptide.

Kallikrein. Proteolytic enzyme that converts kininogens to kinins. *Kalong*: flying fox *Pteropus vampyrus* (Chiroptera: Mammalia) of Malaysia; largest living bat with-wing span of 1.5 metres.

Kangaroo. Large marsupial mammal of Australia having small forelimbs that do not touch ground while running; powerful hind legs that help in Jumping and leaping; females with a pouch in front for carrying the young ones - large grey kangaroo *(Macrops giganteus)* and red kangaroo (M.rujiLs). *Rat:* any small-big rat-sized kangaroo *(Bettongia* spp., *Potorous* spp. of Tasmania: small, pouched. Jumping rodent *(Dipodomys* species) found in deserts of USA.

Kappa-Particles. In some strains of *Paramecium aurelia* individuals have DNA-containing particles which on coming in contact with individuals, not having them, cause their death. These particles do not have a killing effect on the bearer. The individuals bearing them are called killers and the poison produced by them is paramecin.

Karyo. Pertaining to nucleus of a cell or its contents. *Karyocyte*: immature red blood cell.

Karyogamy. Fusion of two nuclei, each derived from two parents during sexual reproduction.

Karyokinesis. Indirect cell division: sequence of events leading to division of nucleus through mitosis/meiosis; may/may not be followed by cytokinesis.

Karyomere. Swollen condition of chromosome seen towards end of nuclear division.

Karyotype. Character of nucleus as described by size, shape and number of mitotic chromosomes.

Katadromous (Catadromous). Fishes that periodically come down to lower parts of river or sea for breeding and spawning - eel etc.

Kea. Large green brush-tongued parrot, *Nestor notabilis* (Psittacifomes:Aves) of New Zealand; feeds on fruits, grubs and carrion but attacks even sheep, when hungly.

Ked. Blood-sucking, Wingless flies (Diptera : Hexapoda), found parastising birds and mammals.

Keratosa. Subclass of Demospongiae (Porifera) having skeleton of matrix made of horny substance-bath-sponges, *Euspongta molltsima, HiPRDspongia lachne* etc.

Kell. Distinct median ridge-likes structure along ventral part of avian sternum.

Keratin. Protein formed by certain epidermal tissues such as vertebrate skin; process of its formation is known as keratinisation.

Kernicterus. Neurological syndrome in which bile pigments are deposited in basal ganglia, may also develop if birth is complicated by a period of hypoxia.

Kestrel. Windhover: small diurnal bird of prey, *Falco tinnunculus* (Aves) of Europe, characterised by its habit of remaining for long time in same place. up in the air.

Ketone. Organic compounds having c=o group *e.g.,* acetone; also formed in the body during metabolic activities and supposed to be important source of energy in some condition.

It is discharged in urine (in form of ketone body) and also through expired air. *Bodies*: in many tissues acetyl coenzymes-A units condense to form aceto-acetyl coenzyme-A. In liver, which contains deacylase, free acetoacetic acid is liberated. This p-keto acid is converted to p-hydroxybutyrtc acid and acetone. Because these compounds are metabolised in liver with difficulty, they diffuse into circulation and appear In urine. The ketone bodies are formed generally during starvation when there is gluconeogenesis.

Kidney. Chief organ for excretion of liquid nitrogen wastes in vertebrates; also often loosely applied to analogous organs in certain other animals as In cray-fish, snails, mussels etc.

Killer-T-Cells. Cellular immunity is mediated by T-Iymphocytes; when these' encounter the' antigens on cells from other individuals they are activated. They enlarge, divide and release lymphoklnins. T-Iymphocytes responsible for lysis of homograft and tumour cells independent of complement are called killer-T-cells.

Killer Whale. Medium sized (10 to 12 metres long). toothed whale, *Orctnus orea* (Odontoceti: Mammalia) having black and white belly and white patch over the eye; feeds mainly on seals but often attacks other large whales.

Kilocalorie. Standard unit of heat energy is calorie which means amount of heat or energy necessary to raise the temperature of 1 gm water by 1 ·C. This unit is also called gram calorie. The unit commonly used in physiology and medicine is Calorie or kilocalorie (Kcal) which equals 1000 calories.

Kinesiology. Study of human motions through anatomical and mechanical principles.

Kinesthetic. Bodily internal cues from receptors in *e.g.*, joints and muscles.

Kinetoblast. Genetically autonomous membrane-bound organelle associated with basal body.

Kinetoplastida. Order of zoomastigophores (Protozoa), colourless, unicellular organisms having one or two flagella in some stage of their life.

King-Fishers. Group of fish-eating birds (Coraciformes: Aves) having long bill and brilliant plumage, catches fish by diving.

King-Herring. Chimaera: small group of fishes (Holocephalii: Pisces), characterised by having features of both skeleton (cartilaginous) and bony fishes: feeds on bottom dwelling crustaceans and molluscs.

King-Penguin. Huge flightless sea-bird. *Aptenodyles patagonica* (Impennre: Aves), found in ice-free seas of sub-antarctic.

King-Vulture. Large brilliantly coloured vulture *Sarcogyps caluus* (Falconiformes: Aves) having sharp, strong, curved beak and powerful feet and claws; native to tropical America.

Kinin. Growth-hormone: substance that induces cell-division.

Kinoplasm. Protoplasm that forms astral rays and spindle fibres. KINORHYNCHA : class of Aschelminthes; microscopic marine spiny-worms with superficial segmentation and without external ciliation - *Centrod.eres* species. *Echinoderes* species.

Kinosternid. Family of Cryptodira (Chelonia: Reptl1l a) having inguinal scent glands that give off musk-like odour - musk turtles.

Kirkbyaacea. Monomorphic superfamily of Beyrichicopina (Paleocopa: Crustacea. comprising of extinct ostracods that were all reticulate.

Kites. Group of falcon-like birds. *Milvus* species (Falconiformes: Aves) dirty brown in colour, curved upper beak. long pointed wings and forked tail; feed on carrion, fish and other small animals.

Kiwi. Smallest flightless bird. *Apteryx autralts* Neomithes: Aves). shy greyish-brown, tailless with long drawn out beak and no wings; good runner. National symbol of New Zealand.

Klipspringer. Small. agile mountain antelope (Artiodactyla: Mammalia). about 50 em high. having very small hooves; males have short. erect homs; found in dry rocky ravines of East and South Africa.

Knee Jerk. Stretch reflex of qudreiceps femoris muscle topping the patellar tendon.

Kneridae. Family of Gonorynchiformes (Pisces), smaliloach-like fresh-water fishes of tropical Africa.

Koala. Largest aboreal bear-like marsupial mammal *Phascolarctos cinereus* of East Australia having thick grey fur. tufted ears and no tail; feeds on eucalyptus leaves mostly during night.

Kookaburra. Laughing jackass: giant kingfisher (Coraciformes: Aves) pattemed with brown and white; makes loud discordant cry. as if it is laughing; found in Australia.

Krait. Common name for deadly poisonous snake *Bungarus* which has large belly scales (crossing lateral. extremities). rounded tail largest 4th infralabial scale and hexagonal dorsals.

Krang. (Kreng). Carcass of whale after removing its blubber. *Krause's end Blulb*: thermoreceptor in skin where skin is hairless: hair follicles of hairy skin and anal mucosa.

Kreb's Cycle. Final oxidation of acetyl coenzyme-A (formed by oxidative conversion of pyruvic acid obtained dUring glycolysis to CO_2 and water. The energy released is used for reduction of certain co-factors mainly NAD and flavoproteins (see glycolysis and ETC).

Krill. Very small (plankstonic) shrimp-like crustacea, *Euphausia* species; main food of large toothless whales.

Kudu. One of the largest African antelops. *Strepsicerus* species (Crtiodactyla: Mammalia);shoulder height upto 1.6 metres, having vertical white stripes on brown body and males with spiral homs 1.4 to 1.5 metres high.

Kupffer's Cells. Liver sinusoids have reticulo-epithelial cells in their walls which are, commonly called histiocytes but in liver particularly, termed the Kupffer's cells. These cells remove foreign matter, dead tissue and bacteria from blood stream by phagocytosis.

L

Labellum. Prolongation of labrum in certain insects specially beetles, bugs and true flies.

Labial. Any structure pertaining to lips like labial artery, labial muscles, labial glands etc. *Glads*: glands found on inner surface of lips in mammals. *Palp:* either of sensitive lobes near mouth of some bivalve molluscs: Jointed appendage attached to labium of certain insects.

Majus. (Pl. Labia Majora); one of the two outer folds surrounding valva of female.

Labrum. Upper lip of insect's mouth: dorsal to mouth's opening. *Laybyrinth*: part of internal ear-membranous labyrinth found in all vertebrates and its hearing portion is called lagena. In mammals the periotic bone also forms bony labyrinth around membranous labyrinth.

Labyrinthine Organs. Special air breathing organs found in climbing fish *Anabas.* Placed in dorsals enlargements of gill chamber. Consist of three or more cocentrically arranged bony plates with frilled margins attached to bony base which in turn is attached to upper end of 4th gill arch.

Lacciferida. Family lac insects (Coccoidea: Homoptera); females apterous, having irregular globular body, degenerate antennre and no legs; males Winged but rare - *Laccifer lacca* and allied species.

Lacertilia. Equivalent name for Sauria, suborder of Squamata (Reptilia: Chordata).

Lacewing. Group of small slender insects (Neuroptera: Hexapoda) with long. knobbed antenna! and large finely veined greenish wings spotted with brown; feed voraciously on other insects.

Lachrymal. In most of vertebrates like birds. mammals etc. large orbit is bounded by a ridge to which the bone attached is called lachrymal. It is also called sometimes supraorbital.

Lactation. Processed of milk secretion and ejection in the mammary glands. It is influenced by the hormone prolactin.

Lacteal. Pertains to milk. Often refers to lymph vessels of small intestine of a vertebrate.

Lactogenic. Milk producing/milk secreting stimulating hormone/ substance like prolactin from pituitary gland.

Lactose. Disaccharide present, in milk only; also called milk sugar. When hydrolysed by lactase. Ii yields glucose and galactose:

Lacuna. Small cavity of space. particularly in cartilage or bone which contains a cartilage or bone cell in life.

Lady-bird Beetle. Group of coccinellid beetles (Coleoptera: Hexapoda), strongly convex. brightly coloured insects with varying number of black spots; both 13.1vae and adults predaceous on aphids. coccids and other small insects and mites .

Laemobathridae. Family of bird lice (Mallophaga: Hexapoda). LAGENA: portion of membranous labyrinth which is concerned with hearing.

Lagomorpha. Order of mammals (Eutheria: Chordata); having two palLS of upper incisors and three premolars above and two below; strongly developed hind-limbs for leaping; long ears but short tail - hares *(Lepus* spp.) and rabbits *(Oryctolagus* spp).

Lamarcksm. Doctrine of organic evolution put forward by Lamarck (1944-1829 based upon (0 effect of environment.

(it) use and disuse, and (itiJ inheritance of acquired characters. The theory was/given in1809 but discarded during 1880-1892. In light of recent findings combination of old views together are referred to as Neo-Lamarckism.

Lamb. Young sheep, *Outs* species (Artiodactyla: Mammalia). *Lamella*: thin leaf-like layer as gill lamella.

LAMINA PROPRIA: middle layer of mucosa of intestine consisting of thin layer of peculiar connective tissue supporting epithelial layer over a basement mambrane and contains· in it small modules of lymphatic tissue, blood vesels and nerves.

Lamprey. Primitive living vertebrate (Cyclostomata: Agntha) long cylindrical body, laterally compressed tail, soft slimy skin without exokeleton: attach itself to large fish, rasping a hole in the tissue of fish and sucking blood therefrom.

Langur. Hanuman monkey (Colobidre: Mammalia), slim primate, about 70 em high with tail longer than body; face, ears and palms black,no cheek pouches but complex, stomach; arboreal and herbivorous - *Presbytts* Spp.

Lantern Fish. common name for deep-sea teleostei fish (Myctophidre: Pisces) characterised by luminous glands widely distributed on body surface.

Lark. Dull coloured perching song bird (Passeriformes: Aves); nests in ground and sings as it flies high - *Alauda gulgula* etc.

Larva. Immature stage in life cycle of various animals which reaches the adult stage by undergoing metamorphosis.

Larvaevorid. Large family of true flies (Cyclorrhapha: Diptera); characterised by thick covering of bristles on body; mostly parasites especially of arthropods - *Echinomyia* Spp.

Laryngel Pouch. Lateral sac-like expansion of cavity of larynx; well developed in certain species of monkeys.

Larynx. Organ situated between trachea and base of tongue. Its opening is called glotte and it contains vocal cords. Found in all vertebrates having lungs except fishes.

Latent Period. Period between occurrence of infection of host by a pathogen and appearance of identifiable symptoms of disease. During this period the activity of pathogen remains suspended. With reference to nerve tissue it refers to isopotential interval following the'; stimules artifact (brief irregular deflection of baseline) when stimulus is applied.

Lateral. Side of the body = at each side of the median line.

Line System. Series of hair cells on medial walls of canals running the length of head and body of fishes and many amphibians. These channels having opening to outside respond to water movements. *Plate Mesoderm*: scizocoelic coelom is derived from mesodenn not originating from enteron. On contrary it arises infonn of a plate-like structures on each side of notochordal process.

L-Dopa (Levo-Dopa). during the biosynthesis of catcholamines dopamine is fonned. L-dopa is isomer involved in this. It is effective in the treatment of Parkinson's disease and involved in the control of prolactin secretion in animals and human beings.

Leaf-cutting. Bee small. solitary bee. *Megachile* spp. (Hymenoptera: Hexapoda) having stout, densely haired body, nests in cavities of wood and stems, that are lined from inside with cut portions of leaves.

Leaf Hopper. Small elongated hopping insect (Cicadellidce: Hemiptera) having sucking mouthparts; feed gregariously on all kind of Yegetation. some act as vectors of various plant diseases.

Leaf-Insect. Leaf-like Winged insect *Phyllium crucifolium* (Phasmida: Hexapoda) with flattened body, fore-wings modified look like a leaf. hind wings membranous.

Learning. Ability of nervous system of an individual toward an extemal stimulus in a particular way based upon some past experience.

Leather Jacket. Brown thick-skinned legless larva of crane-fly (Tipulidce: Diptera), feeds on grass roots. *Turtle*: largest sea-turtle (Chelonia: Reptilia) abo,ut 2 metres long with large fore-flappers; shell made up of numerous bony plates covered with leathery skin; found mostly in warmer seas.

Lectotype. Specimen selected from original material, after publication of its description; to serve as type specimen.

Leech. Aquatici bloodsucking annelids (Hirudinea).

Lemming. Small rodent of Arctic and subarctic regions reputed in migration to rush headlong into sea and get drowned - *Lemmus lemmus* (Muridce).

Lemniscal System. Neurons from gracile and cuneate nuclei across mid line ascend in medial lemniscus to end in specific sensory relay nuclei of thalamus from dorsal column or leinniseal system,

Lemur. Group of mammals akin to monkey; but with pointea muzzle: several *Lemur* species (Lemuroidea: Primates) 15 em to one metrif long and an equally long bushy tail; forest-dweller, nocturnal, confined to Malagasy.

Lens. Transparent structure in the eye having refractive power and derived from ectoderm of head.

Leopard. Large, ferocious, terrestrial animal, Panthera *(Felis) pardus* (Carnivora: Mammlia), tawny-yellow body with black spots, unguiculate with strong claws; found in Southern Asia, and Africa; feeds on other higher animals and will not spare even human beings.

Lepidoptera. Second biggest order of insects (Hexapoda), Long antennre, suctorial proboscis, membranous Wings; body, wings and appendages covered with overlapping pigmented scales that give the species its colour pattern; metamorphosis complete -butterflies moths and skippers.

Lepidosauria. Subclass of Diaspida (Reptl1la), skull having two temporal openings on each side with reduced bony arcades -Eosuchia, Rhynchocephalia and Squamata

Leptocardii. Claas of Cephalochordata (Acraniata Chordata) slender fish-like animals having one-layered epidermis. many gill-slits and no scales - lancelets.

Leptostraca. Primitive small marine crustaceans (Malacostraca) having large carapace held by adductor muscles; eyes stalked; telson bearing two movable articulated prongs.

Lethal Gene. Mutant allele that kills the bearer (Individual possessing it). The mutation commonly takes the form of a small chromosomal deletion so that genetic code for a key position is missing or dissrupted - yellow lethal in mice.

Leucine. Neutral amino-acid; amino-acid with unsubstituted chains. Essential amino-acid. Its isomer is isoleucine which is also an essential amino-acid.

```
          COOH
           |
   NH2 – C – H
           |
          CH2
           |
   CH3 – CH
           |
          CH3
          COOH
           |
   NH2 - CH – H
           |
   CH3 – CH
           |
          CH2
           |
          CH3
```

Lecuine and Isoleucine

Leukemia. Abnormal increase in WBCs number from 7000 to 700,000 (or even more). Considered a type of blood cancer.

Levator. Muscle which raises a lxx:lily part.

Leydig Cells. Cells of testes that are stimulated by luteinsing hormone to secrete testesterone

Life-Cycle. Definite pattern of birth, growth, reproduction, old age and death.

Ligament. Tough. fibrous connective tissue band that connects bones of supporting viscera.

Limbic System. In mammals and man the small portion of rhinencephalon is concerned with olfactory power while rest of it is concerned with emotional responses, instincts and complex neutroregulatory functions and is termed limbic system or Umbic lobe. Each limbic lobe consists of a rim of cortical tissue around hilus of cerebral hemisphere and a group of associated structures - the amygdala, hippocampus and septal nuclei:

Limbs. Paired locomotory structures of tetrapods: There are two pairs of limbs - fore limbs and hind limbs. Each limb consists of three parts- brachium or stylopodium forming upper arm in fore limb and thigh in hind limb: *antebrachium* or zeugopodium forming fore arm in fore limb and shank in hind limb; autopodigm forming hand in fore limb and foot in hind limb. The autopodium is further divisible into basipodium forming wrist in fore limb and ankle in hind limb, metapodium forming palm in fore limb and sole in hind limb. acropodium forming fingers in hand and toes in foot.

Limnaeid. Fresh water or pond-snails (Limnaedre : Mollusca).

Limnology. Scientific study of fresh-waters.

Limnoplankton. Plankton found in fresh waters such as lakes. ponds and pools.

Limy. Containing Ca++ salts particularly CaC03.

Lingual. Tongue-lingual artery. lingual vein etc.

Linkage. Association of genes on same chromosomes. Genes that occupy loci which are close together will tend to be transmitted together. Those which are found farther apart may be separated frequently by crossing over events making the demonstration of linkage difficult through genetic analysis.

Linked Gens. Genes which are associated on same chromosome and thus tend to be transmitted together in a cross. Since linked genes occur at different loci. they are not alleUc.

Limpet. Gasteropod mollusc with low conical shell that clings tightly to the rocks and is not easily separatable-Patella spp.

Linnet. Shy finch, passertne song-bird, *Carduelts cannabina* (Aves) of Old World, brown or warm-grey in colour; nesting in gorse and dense bushes.

Linsang. Slender. mangoose-like, nocturnal mammal of South-east Asia having brownish-grey short velvety fur with dark spots or bands.

Lion (Lioness). King of beast Oungle) huge carnivorous mammal *Panthera (Felis)* Leo (Felidre) confined to South of Sahara in Mrtca and Saurashtra in India, having tanny body with black-tufted tail; lives in groups (prides) and hunt large herds of antelope and zebra.

Lip. Fleshy fold above and below the entrance of mouth of mammals: edge or rim of cavity, orifice, vessel or wound.

Lipase. Fat digesting enzyme found in intestinal Juice. Also reported to be present in gastric secretion. It selves to convert fats to fatty-acids and glycerol.

Lipemia. Milky appearance of plasma after meals due to presence of so many large particles in blood which are called chylomcrons and are formed in the intestinal mucosa during absorption of long chain fatty -acids into lymphatic ducts. The lipemia is cleared by action of lipoprotein lipase.

Lipogenesis. Formation of fat from non-fatty sources in the body of individual.

Lipotropins. Peptides with fat-metabolising activity have been isolated from pituitary gland of mammals *e.g.,* sheep. g-lipotropin contains 58 amino acid residues while b-lipotropin may be precursor of peptides that bind to opiate receptors (enkephalins). Both lipotropins have MSH (melanophore. stimulating hormone) activity in the body.

Litoral. Floor of sea from shore to the edge of continental shelf. **LIVER:** one of the exocrine glands associated with digestive system secreting mainly bile juice and concerned with a number of functions related to metabolism of body like sugar control, fat control, storage and metabolism of essential products. formation of ammonia, urea, uric acid etc., detoxification of harmful substances, elimination of foreign and dead materials, hydrostasis etc.

Lizard. Quadruped reptile (Lacertilia: Squamata) having non-expansible mouth, movable eyelids and cardrums and fleshy-tongue snake-lizard is limbless and burrowring type.

Llama. South American mammal *Lama glama* (Camelidare Artiodactyla) camel-like, but smaller, humpless and woolly-haired; used as beast of burden.

Lobi Inferioris. Hypothalamus in fishes has a pair of accessary lobes that are concerned with smell and taste.

Lobster. Marine arthropod, *Pariulirus* Spp (Crustacea) which lives in rocky regions of sea; bilaterally symmetrical and segmented body comprising cephalothorax and abdomen; five pairs of walking legs and pleopods and tail-fin with telson to help in swimming.

Locus. Specific position occupied by a particular gene on a chromosome. Anyone of the alternative forms of genes may be formed at a given locus.

Locust. Most destructive phytophagous insect (Acrididre : Orthoptera) - Bombay Locust, desert Locust and migratory locust are commonly found in India.

Loon. Fish-eating, aquatic diving bird, *Gavta* spp. (Gaviiformes: Aves). **LOOP OF HENLE:** kidney structure found in the medulla in case of mammals where most of water reabsorption takes place. Its activity is controlled by ADH (antidiuretic hormone). It is concerned with diuresis (Urine formation).

Lophodont Teeth. Molars in elephant and horse have cusps fused by means of intermediat.e. masses of dentine to form ridges or lophs.

Lophogastrida. Suborder of free swimming marine crustaceans (Malacostraca), carapace covers the thorax; gills seven pairs, tail-fin well developed.

LDS (Lysergic Acid Diethylhmide). Psychomimetic agent and serotonin antagonist. The compound produces transient hallucinations and other mental aberrations.

Luciferase. Enzyme that contributes to light production in certain animals like fire-files.

Lumen. Internal cavity within body or a structure like lumen of intestine, gland or a blood vessel.

Luminiscence. Production of light as a result of chemical reactions in the cells as production of bluish-green light by Noctiluca (a marine protozoa) during night hours.

Lumpsucker. Clumsy, spiny-finned, dark-coloured, sluggish fish Cyclopterus lumpus having ventral sucker or adhesive disc on belly with whichit clings to seaweed.

Lungs. Respiratary organs in air-breathing vertebrates. *Book.* breathing organ of scorpions and spiders. *Fishes*: air-breathingfish that has both gills and lungs (Dipnoi: Osteichthyes): with help of lung, it engulfs air to perform respiratioin; foudn only intropical rivers.

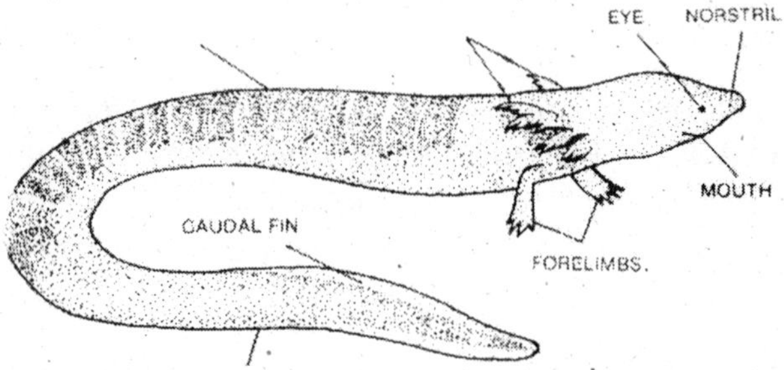

Luteolysis. Regession of corpus luteum so that next menstrual cycle may begin.

Luxation. Easy passage of faecal matter out of alimentary canal. Gastrointestinal motality. mucous secretion by large intestine contribute in the process.

Lychniscosa. Order of Hexasterophora (Porifera), that inrlllc1es sponges in which parenchymal megascleres form a rigid framework.

Lymphagogues. Agents which increase lymph flow and include a variety of agents that increase capillary permeability and the agents causing contraction of smooth muscles.

Lymphocytes. Agranulocytes (type of WBCs) with round or kidney shaped nucleus. Constitute about 28% of WBCs. Produce antibodies and opsonins to destroy foreign matter.

Lympyridae. Family of glow-worms (Cantharoidea : Coleoptera) having phosphorescent organs on underside of hind abdominal segments, these emit light.

Lyon-hypothesis. Concept that in any given cell of a mammalian female one X-chromosome is entirely active. the other being inactive completely or in parts. The inactivation occurs at random so that the patemally derived X-chromosome is active in some cells and maternally derived one active in others.

Lysis. Break-down of organised structure of biological matter; usually restricted to break-down of living cells.

Lysolecithin. Lipase-A in the pancreatic juice splits a fatty acid off lecithin' forming lysolecithin. This substance damages cell membranes. In acute pancreatitis. a severe disease. there is activation of phospholipase with the formation of lysoloecithin from lecithin. This causes disruption of pancreatic tissue and surrounding fat.

M

Macaw. Large, brightly coloured. long-tailed parrots (Psittaciformes: Aves) of central and South America, about 60 cm long with hooked beaks. live in flocks and feed on fruits and seeds.

Macergate. Large-sized worker ant (Formicidre : Hymenoptera).

Mackerel. Edible sea-fish *SComber scombrus* having greenish back with dark blue stripes and silvery belly. swim in large shoals near surface. *Shark* : blue dog or porbeagle *(Lamna nasus):* common name for various pelegic predacious fishes (Isuridae : Selachii) having heavy body. sharp awllike teeth and symmetrical tail - Isurus *glaucus* (Mak of Pacific-Indian oceans).

Macrocephalia. Large-headed: person having head with cranial capacity above 1450 cc.

Macromere. Relatively large yolk laden cell that is present during cleavage stages of certain animals like frog etc.

Macronucleus. Large nucleus in ciliate - protozoa *Paramecium* spp., controlling all body activities except reproduction.

Macroparasite. Parasite that can be seen with naked eye-louse.

Malingant Transformation. Change from non-cancerous to cancerous state.

Mallard. Common wild duck, *Anas boscas* (Anseriformes: Aves) ; male (drake) is brightly coloured with glossy green head and neck, brown breast separated by white ring; female dull coloured with mottle-brown plumage.

Mallus. First ear ossicle attached to tympanum in the ear of vertebrates.

Mallophaga. Order of wingless paraneopteran insects (Hexapoda), biting and chewing mouthparts, eyes reduced, ocelli absent. cerci wanting. metamorposis gradual ; mainly ectoparasites on birds - bird lice.

Malpighian Body. Renal Corpuscle structure in kidney of verterbrates consisting of Bowman's capsule and glomerulus.

Maltese Cross. Centrum in carttlagenous fishes is not purely cartilagenous; calification takes place and fibrocartilage develops in whole centrum except at 4 wedge-shaped places which are called maltese cross.

Mamba. Large. deadly-poisonous African snakes *Dendraspis* spp. (Ophidia: Reptilia) - black mamba. green mamba.

Mamma. Structure in skin of mammals where various mammary glands open through their ducts; characteristic shape and bulk of the structure is formed by adipose tissue surrounding mammary gland. The mammae open at niples formed by erectile tissue: distribution is generally pectoral (woman) or inguinal (cattle) or from pectoral to inguinal (pig.).

Mammalia. Group of warm blooded animals with hair on skin and mammary glands producing milk which is used to nourish young ones.

Mammalogist. Zoologist who has sperialised in mammalogy. *Mammalogy*: branch of zoology dealing with study of mammals.

Mammary Glands. Modified sweat glands of skin in mammals haVing milky secretion; compound saccular and apocrine type in structure (except· egg-laying mammals where these are compound tubular). Each has many lobules and exceretory ducts from lobules unite to form large ducts which meet similar ducts and pour into main ducts which open out separately at the nipple.

Manatee. Sea-cow: seat like aquatic manimal (*Sirenia* : Eutheria) grows upto 5 metres in length, has thick blubber and no hair; found in warmer seas; feeds on aquatic vegetation.

Mantle. Folding of the body enclosing soft structures of an animal. *e.g.*, molluscs. which secretes shell. Also soft covering on the body of urochordates like *Herdmania.* spp.

Manubrium. Structure projecting from the middle of subumbrellar surface of medusa of enidarians and bearing mouth at its free end.

Manus. Part of forelimb of tetrapod animals corresponding to hand of human being.

Map Unit. Figure which measures distance between two linked loci as inferred from the amount of crossing over which occurs between them. For loci which are close to each other, map units are equal to number of new combinations which can be detected so that 1 map unit = 1 % recobinations. For loci more distantly situated with reference to each other, the amount of multiple cross even cuts down the number of new combinations which can be detected so that the map units and 1 % recobinations are not equivalent.

Marine. Of sea-water; forms or bodies inhabiting salt-water.

Marine Biology. Branch of biological science that deals with marine organlsms (animals/plants).

Marker Gene. Any specific gene or genetic trait which can be used to follow the transmission of a specific chromosome or chromosomes region in mating.

Marrow. Tissue filed in the spaces of long bones like femur etc. The marrow of vertebrates in general is yellow bone marrow which stores fat but in mammals it is red bone marrow besides yellow bone. The red bone marrow has four types of cells -haemocytoblasts which form RBCs, myeloblasts which form WBCs (granulated), monoblasts that form WBCs (granulated) and megakaryoblasts which form marrow platelets.

Marsupium. Pouch on ventral side of mother in some mammals *e.g., Macropus* (kangaroo) where rest of development of premature (precareous) embryo occurs. Nipples are located inside the pouch for nourishing young ones with milk.

Masseters. Muscles behind temporalis (arising from otic capsule and inserted on coracoid process of lower jaw) arise from quadratojugal to be inserted on outer surface of lower jaw.

Mass Pyramid. Mass relationships within an ecosystem. Moving towards apex, mass decreases *e.g.,* base may be composed of producers while apex has secondary or tertiary consumers.

Mast Cells. Histiocyte-like cells in connective tissue around the blood vessels is in the areolar tissue. These secrete an anticoagulant Heparin.

Masticatioin. Chewing food with teeth or other mouth parts as in many insects and other animals like mammals.

Matigonemes. Small delicate processes arising from one side of flagellumas in *Euglena* species or similar type of animals.

Mastigophora. Group of ptotozoa a bearing flagluum *e.g., Euglena* species usually autotrophic bearing chlorophyll.

Matodon. Extinct elephants of America, *Mammut* species (Proboscidea: Mammalia) that lived about ten thousand years ago and roamed from Alaskato Mexico ; more or less same size as present day elephants but body was covered with long coarse hair and they had -nipple-shaped tubercles on crown of molar teeth.

Mating. Meeting or coming together of individuals of opposite sex, for sexual intercourse, may be for purpose of pleasure or reproduction. *Strains* individuals of a species that do not interbreed with each other but are able to breed with other individuals of same species that are morphologically identical but genetically different.

Maxilla. One of the several appendages of mouthparts of various arthropods. A large bone of upper Jaw in the gnathostomes (vertebrates).

Mechanoreceptors. Sense organs responding to mechanical distortion or pressure.

Mecoptera. Order of carnivorous insects (Hexapoda); head produced into deflected rostrum, antennce long and filifonn, mouthparts biting and chewing; wings long, narrow subequal and membromous; last abdominal segment raised upward like that of scorpion; metamorphosis complete - scorpion-flies etc.

Mediastinum. Thoracic cavities in mammals are fonned of two membranes - outer parietal membrane lining the thoracic walls and inner pleural membrane surrounding the lung. Parietal membranes of two sides become opposed to ea.ch other in median line to form a median septum called ~ mediastium enclosing mediasinal space (that encloses the heart).

Medulla. Inner region of a gland or other structure in animals. In endocrinology refers to inner portion of adrenal gland secreting catecholamines.

Medullary Groove/tube. Neural plate groove or tube: successive stages in embryonic development of vertebrate central nervous system.

Medulllary Sheath. Sheath around the medullated fibres; made up. of faity substance between which there are thinner layers of protein. Also called myelin sheath.

Megatherium. Gigantic extinct ground-sloth of South, America (Edentata: Mammalia).

Meiosis. Last stages in the development of gametes characterised by two divisions in which number of chromosomes is reduced to half (to hapliod number).

Melanoblasts. Branched cell found in the skin of mammals controlling skin pigmentatlon. Branches twine around the epidermal cells in the neighbourhood and teminate in cup-shaped endings around an epidermal cell Produce melanocytes.

Melanocytes. Cells in the skin of mammals bearing melanophores (produced from melanoblasts).

Melatonin. Indole derivative found in pineal gland of mammals and other vertebrates. It is seen to lighten the skin colour of tadpole through exertion on melanophores. But is mammals the function of the substance remains unestablished. There is speculation that it exerts anti gonadotropic efect but the effect is destroyed by light.

Menorrhagia. Abnornally profuse bleeding during regular menstrual cycles.

Menstrual Cycle. A cycle found in female of mammals belonging to primate group (apes and men); composed of four phases and completed in about 28-30 days in majority of cases. Its various phases are:

Menstrual Phase. Lasts for 4-5 days and there is discharge of mucous and blood from uterine wall. *Proliferative Phase*: period between 5-13 days during which uterine wall is repaired and ovum contained in Graafian follicle matures. *Ovulatoryphase*: period of 14 to 16 days during which ovum is released in oviduct and fertilisation likely to occur.

Menstruation. Discharge of blood and uterine tissue in female human being and apes at the end of menstrual cycle in which fertilisation did not occur. Also called menses.

Mental Glands. Tubular glands below the chin in some male salamanders; develop as a secondary sexual character. Volatile exudations from these attract females.

Mentomeckelian. Bone replacing the tip of meckel's cartilage in the lower Jaw of frog.

Merocrine. Glands in which secretory cells discharge their products without causing much damage to the cells and after discharge the cell undergo a period of rest during which they are seen to undergo refilling *e.g.*, mucou,s and poison glands of amphibia.

Merozoites. Mature trophozoites of sporozoa like. *Plasmodium* after mitotic division.

Metacromian Process. Backward process arising from acromial process of mammals developed specially in cases which have reduced clavicle.

Metamerism. Serial repetition of homologous organs as found in arthropods. annelids and chordates.

Metamorphosis. All those changes that bring about development of larval stage to adult. In other sense marked structural changes or transformation during development as from larva to adult.

Metanephridwm. Tubular excretory organ of excretion as found in earthworm leading to outside through coelom.

Metanephros. A kidney found in amniotes (reptiles. birds and mammals) derived from posteriorly originated tubules of embryo.

Metapleural Folds. Two folds on the anterior central surface of *Amphioxus* Species extending between mouth and antriopore.

Metasoma. Posterior region of abdomen of certain arachnids. as in scorpions etc.

Metazoa. Multicellular animals with tissue grade of organisation *e.g.*. all animals beyond sponges.

Metazoea. Late larval stage in development of certain crustacea such as crabs etc. showing presence of carapace and rudiments of thoracic limbs.

Metecdysis. Short period after each moult in the arthropods. during which the new cuticle hardens and the animal becomes normal.

Micromers. Very small cells formed by unequal 3rd cleavage during the early development. These cells are destined to form ectoderm of adult.

Micro-organism. Minute organism. visible only under microscope and not by naked eye.

Micropenis (microphallus). Very small penis (phallus)

Micropodiformes. Order of very small birds (Carnatae : Aves) that are very good fliers. having long wings -humming-birds. swifts etc.

Micropyle. Minute opening in egg shell through which spermatozoon may enter ovum.

Microscope. Optical instrument for producing magnified image of small objects by using single or combination of lenses - simple. compound. stereoscopic. electron. ultra-violet etc.

Microscopist. Person expert in handling and using various microscopes.

Microscopy. Study of minute objects with help of microscopes.

Microtome. Apparatus for cutting extremely thin sections of material for microscopy.

Microvilli. Tiny/minute folds at the surface of cells *e.g*., in the intestinal columnar cells.

Midge. Common name for various small. delicate. mosquito-like dipteran flies (Oligoneura : Nematocera) belonging to three families. Cecidomyiida!. Ceratopogonida! and Chironomida! -phytophagous and parasitic

Midgut Diverticulum. In animals like *Amphioxus* the central projection from the midgut .

Migration. Seasonal movement of large number of birds as a habit to obtain advantage of favourable conditions offered in other regions mainly for exploring new areas for nesting.

Millepore. Hydrozoan coral with many pores or polyp-cells.

Millipede. Thousand-legged (exaggeration); small herbivorous arthropod (Diplopoda : Myriapoda); body cylindrical, head with one pair of antenna!. most of body segments (except first three) bear two pairs of jointed legs; breathe by air-tubes.

Mimicry. Adaptive resmnblance of certain animals to other surroundings. resulting in protection from predators *e.g.*, frog when sits in green grass changes body colour in same pattern and becomes undetectable to enemies.

Moa. Large, recently extinct flightless bird, *Dinomis maximus* (Dinornithidae : Aves) of New Zealand, resembling ostrich, 2.5 to 2.7 metres tall and weighing up to 2.5 qUintals.

Modifier Gene. Individual carrying a pair of alternative gene forms (alleles) . heterozygous at a single locus under consideration. Also a cross in which a single pair of alleles is being followed.

Molars. Grinding teeth in heterodont dentition; Multirooted with broad crown having ridges, furrows, grooves, cusps etc. for grinding food: permanent teeth.

Mole. Abnormal pigmented prominence on skin: small burrowing insectivore mammal, *Talpa micrura* has flattened body short blackish-grey velvety fur; pointed snout, rudimentary eyes covered with skin and wide clawed feet; feeds on worms. *Crab.* burrowing crustacccan(Mollusca) of sandy beaches, intermediate between crab and lobster having short, flat fore-limbs modified for digging and scooping. *Rat*: burrowing rodent, *Nesolia indica* (Rodentia: Mammalia). SHREW : common name for several burrowing insectivores, intermediate between moles and shrews, 12 to 15 cm long having long snout and very small ears and eyes - musk- mole-shrew, *Crocidura murtna* frequents houses at night in search of food (insects etc.) and utters sharp, shrill cry. *Skin*: soft fur of mole, cropped before dyeing.

Mollusca. Non-segmented, schizocoelomate phylum comprising of terrestrial, fresh-water and marine forms descendents of ancient marine polychaete annelids -mussels, snails, slugs, octopus etc.

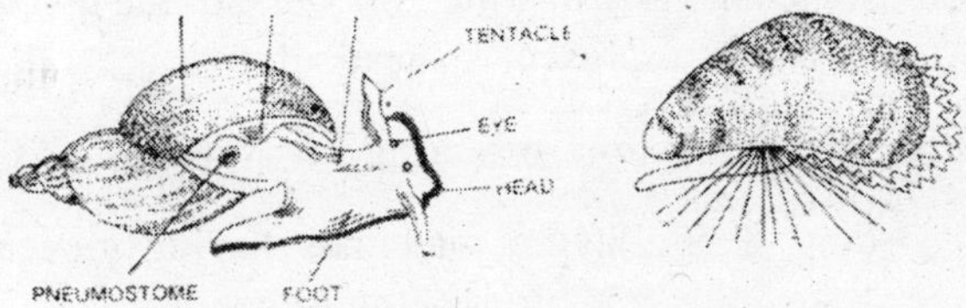

Mollusca

Moner. Major division of organisms characterised by absence of nuclei and chromosomes; include some bacteria and blue green algae.

Mongoose. Small carnivore, *Herpestes* spp. (Fissipedia: Mammalia), related to civets and genets, 50 to 60 em long body, sharp muzzle, short legs and long bushy tail another 50 to 60 em, very alert and active, known for its ability to kill snakes, including poisonous ones.

Monimostylic Jaw. Autostylic Jaw suspension where quadrate is immovable: found in lizards.

Monkey. Common name for most primate mammals except apes and man.

Monkfish. Ray-like shark (Pisces), grows upto 2.5 metres in length, bor.tom-dweller in many subtropical and temperate seas; feeds on I1atfish and other small animals of seabed.

Monoecius. Hermaphrodite: both male and female reproductive organs present in a single individual: opposite to dioecious.

Monomorphic. Existing in only one form.

Monomyarian. Having only one adductor muscle.

Monophyodont Teeth. Single set of teeth in life as molars of mammals: teeth that are not replaced.

Monosaccharide. Simple sugar which can not be further hydrolysed glucos.e, fructose, galactose. pentose etc.

Monsomy. Chromosome anamoly in which only one of the given kind of chromosomes is present instead of two.

Monotocous. Producing only one offspring at a time.

Monotremata. Group of egg-laying mammals having cloaca instead of separate mius and urinogenital openings.

Moropus. Miocene epochossil. mammal akin to horse and rhinoceros but had claws instead of hooves; shoulder height about 2.5 metres: fed on tree leaves in forests of North America; with disappearance of the forests these beasts also became extinct.

Morphology. Branch of science related to study of form and external structure of an organism.

Morula. Mass of cells (64-128) formed by division of zygote during the development ormost animals.

Mosaicism. Occurrence of two or more genetically different cell types within the body of an individual.

Mosasaur. Huge eel-like aquatic reptile that lived during Cretaceous period and was known as king of sea; slim. streamlined body. about 12 metres long. with paddle- shaped limbs. predaceous on fish.

Motor Neurons. Neurons which have the cell bodies lying in the gray matter of CNS and their fibres always leaving via ventral roots. They inervate.the effectors.

Moulting. Casting off exoskeleton. to shed portions of skin. feathers or hair.

MSH (*Melanocyte/Melanophore Stimulating Hormone*). Peptide hormone released by adanohypophysis that affects melanin distribution in mammals and certain skin-coloured changes in lower vertebrates like fishes. amphibia and reptiles.

Mucin. Glycoprotein (protein conjugated with glucose) which lubricates food. It is found in saliva. gastric secretion and large intestinal secretion.

Mucosa. Embrane that secretes mucous. Usually restricted to innermost lining of certain organs like divisions of alimentary canal and uterus etc.

Mucous Glands. ('glands in skin. mucosa of alimentary canal secreting mucopoly-saccharide mucin which is a slimy structure.

Mucron. In some sporozoa *(Monocystts* etc.) anterior structure in some individuals which is modification of a portion of ectoplasm.

Mullerian Ducts. Paried embryonic ducts originating from peritoneum that connect with urinogential sinus to become oviduct, uterus etc. in adult.

Muller's Organ. Wheel organ: ciliated finger-like projections or patches on the undersurface or oral hood of *Amphioxus;*

sets up water current that brings food inside the oral hood.

Multiple Alleles. Three or more gene forms associated with a given locus. Anyone form can occur at a locus but being diploid. any individual can carry not more than two of members of a multiple alletic series.

Multiple Factor Inheritance. Inheritance involving many genes at different gene loci and the complex interactions of many unknown environmental factors .

Multiple Fission. Asexual reproduction in which nucleus performs several mitotic divisions before any cytokinesis

Mumps. Pathogenic condition leading to inflamation of parotid glands.

Muscone. Substance that is reponsible for characteristic odour in musk. obtained from musk-deer *(Moschus moschiferus)* from a pouch under skin of abdomen of male.

Muscles. Connective tissue articulating bones with each other, giving rise to a definite shape to body and helping in locomotion and movements of body parts.

Muscular Dystrophy. Vitamin E deficiency, may lead to reduced and disturbed functioning of body muscles in most of animals. not reported in man. *Twitch*: contraction offered by skeletal muscle with response to single stimulus. It has latent period of 0.1 second, contraction period of 0.4 second and relaxation period of 0.5 second. Twitch may be as long as 60 m1lliseconds as in large cells and as small as10 milliseconds as in muscles of eye-ball.

Musk. Odoriferous reddish-brown substance secreted by male musk-deer; used in perfumes. *Beetle*: longicom beetle (Cerambycidre: ColeOptera) that smells like roses. *Deer*: small hornless deer *Moschus moschiferus* (Artiodactyla: *Mammalia*) found in Himalayas and Central Asia; musk-bag. a.. hairy globular sac about 4 cm in diameter lies at posterior end of penis.

Mutation. Change in a gene that results in a new hereditary variation (trait). Broadly used to include all kinds of hereditary variations resulting from gross chromosome changes. Mutations may also occur in somatic cells. but these are not inherited (unless individual reproduces asexually). *Pressure*: spontaneous rate of gene mutation which occurs continuously in a population.

Mycetophagidae. Family of fungus-gnats (CucuJoidea: Coleoptera) serious pest of mushrooms.

Myelencephalon. Remaining hind brain of embryonic neural tube (rhombencephalon) after metaencephalon (from which arises cerebellum) that gives rise to medulla oblongata (in adult).

Myelon. Remaining part of neural tube after pro-. mesen- and metaencephalon that gives rise to spinal corel.

Mymaridae. Family of very minute insects (Chalcidoldea: Hymenoptera) that are exclusively egg parasites-fairy- flies.

Myocardial Infraction. Serve reduction in blood supply to a portion of myocardium due to arteriosclerosis. thrombus formation or other physiological disorder leading to the death of myocardial cells and formation of necrotic area.

Myocommata. Connective tissue partitions between two successive myotomes as seen in *Amphioxus* and fishes etc.

Myogenic Heart. Heart whose activity is not directly controlled by nerves but is controlled directiy by muscles *e.g.*, hearts of all vertebrates. Compare with neurogenic heart.

Myoglobin. Oxygen binding pigment in red or slow muscles and in respiratory enzyme cytochrome-C. contains haeme in the structure.

Myopia. Short sightedness when eye-ball is relatively short leading to eye's point of focus lying behind instead of on retina.

Myopsida. Group of cephalopod molluscs (Decapoda); eye covered by skin-of head.

Myotomes. Dorsal parts of embryonic mesodermal pouch which gives rise to muscles. More clearly seen in *Amphiauls.* fishes etc. where these look like segmental muscle blocks arranged on each side of body.

Myotonic Dystrophy. Disorder characterised by progressive weakness of muscles. develpment of catract and also cardiac abnormality associated with a dominant autosomal gene.

Myriapoda. Class of Arthropoda that includes millipedes (Diplopoda), centipedes (Chilopoda) and Pauropus) (Pauropoda).

Myxobacteria. Small group of rod-shaped bacteria characterised. by gliding movement and having delicate, flexible cell wall.

Myxoedema. Thyroid deficiency disease that causes swelling within and beneath the skin.

N

Nacreous Layer. Iridescent innermost or pearly layer of some mollusc-shell, consisting of calcium carbonate secreted by epithelium of mantle.

Nacrolepsy. Disease in which there is an eventually irresitible urge to sleep during day time activities

NAD (*Nicotinamide Adenine Dinucleotide*). Co-enzyme serving as hydrogen acceptor; substance picks up hydrogen forming NADH (Dihydro-Nicotinamide Adenime Dinucleotide).

NADP. Another coenzyme which serves as hydrogen acceptor and gets changed to NADPH (dihydro-nicotinamide adenine dinucleotide phosphate) during electron transport system.

Nails. Characteristic structures found on the digits in primates like man; have nail plate and subunguis very much reduced and represented by a narrow strip of less modified epidermis below the edge of the unguis. Dermis of nail bed underlying the unguis is highly vascular as evidenced from pinkish colour of nail.

Nany. She-goat. Capra species (Artiodactyla: Mammalia). NAPE: back of neck.

Nares. Nostrils: internal and external openings of nasal cavity.

Narwhal. Medium sized arctic whale, *Monodon monoce11lS* (Cetacea: Mammalia), grey with darker spots. grows up to five metres in length. male has one (rarely two) long, straight, spirally twisted tusk, about two metres long: feeds on cattlefish. squids and other molluscs.

Nasolacrymal Duct. Duct from lacrimal sinus and terminating into sinuses inside certain bones of nasal region which in turn drain into nasal passage.

Nasopharynx. Upper (nasal) part of pharynx in mammals separated from lower part (oropharynx) by soft palate.

Natality. Birth-rate; frequency or number of births in proportion to population.

Natantia. Suborder of Decapoda (Eucardia: Crustacea); having long rostrum aP.d ventrally fixed abdomen - shrimps etc.

Natural Pathenogenesis. Development of an egg without fertilisation during normal life-cycle of an animal as in aphids (insects). *Selection*: process whereby the environment acting on phenotype indirectly determines which genotype will be represented in succeeding gene pool, hence it has power to alter gene frequencies.

Nautilus. Primitive cephalopod mollusc; having chambered spiral shell and living in terminal or largest compartment. other compartments being filled with gas that gives the shell buoyancy.

Navicular. Boat-shaped bone; one of seven tarsal bones in foot situated on inner side of ankle.

Neanderthal. Man extinct man Homo *sapiens neanderthalensts* which had protruding face and receding jaws; lived in caves and wore animal skins.

Neck. Part of body (usually narrow) that joins head with trunk of vertebrates.

Necrophily. Morbid liking for being in presence of dead-bodies. NECROPHOBIA: aversion or horror to corpses.

Nectocalyx (*Nectophore*). Bell-shaped aquatic polyp having no manubrium or tentacles.

Nectonematoidea. Monofamtltal and monogeneric order of

Nematomorpha (Aschelminthesl, marine forms; characterised by dorsal and ventral epidermal chords, rows of bristles and a pseudocoele; start life as crustacean (crab etc.) parasite and later become free-living (swimming).

Neddy. Ass or donkey, *EqWJS* spp. (Perissodactyla: Mammalia). NEEDLE-FISH: garpike or pipe-fish *Leptsosteus* spp. (Actinopterygii: Pisces); extremely thin cylinderical body. silvery coloured and having narrow sharp-toothed jaws: preys on other smaller fishes.

NEFA (*Non-essential Fatty-Acids*). Free fatty acids, which are present in plasma and are bound to lipoprotelnic complexes. These do not include linoleic acid. linoleinic acid and archdonic acid which are essential fatty-acids.

Nekton (*Necton*). Assemblage of actively swimming organisms in lake. sea or ocean.

Nelly. Large sea-bird. *Macronectes giganteus*. the giant petrel (Procellariiformes: Aves).

Nematocide. To xicant effective In killing nematodes. *Nematocysts*: stinging capsules found in *Hydra* like animals (Coelenterates); each produced by single cell called cnidoblast.

Nematoda. Newly erected phylum that includes elongated cylinderical worms with mouth and anus at two opposite ends; digestive tube separate from body wall by fluid-filled cavity; found in soll. water and as parasites of animals and plants -hookworms. roundworms etc.

Nematoda. Microscopic. thread-like unsegmented round worm (Nematode) both free-living and parasite on plants and animals.

Nematologist. Person having sound knowledge of Nematology.

Nematology. Branch of zoology confined to study of Nematodes.

Nematomorphs. Group of animal. like hair-snakes with thread-like body, blunt anterior end rough cuticle. adults free living but larvae parasite in Insects. *e.g., Cordius.*

Nemertea. Phylum of bilaterally symmetrical ribbon-live unsegmented worms (Acoelomata) elongated body having ciliated epidermis. no body cavity; complete digestive tract with mouth and anus; mostly oviparous - bootlace worms.

Neocortex. Part of cerebral hemisphere which receives information directly or indirectly from all receptors in the body and each muscle and gland cell Is Influenced by impulses originating in it. It is arranged in six layers numbered I-VI from without Inwards. *LayerI*: molecular layer containing numerous dendrites. axons and glial cells. *LayerII*: external granular layer consisting of small densely packed cells which are rounded and polygonal. *Layer* IV: internal granular layer resembling layer II. *Layer* V: internal pyramidal layer consisting typically of cells which resemble those found in layer III. *Layer* VI: fusiform cell layer consisting of long spindle cells arranged perpendicularly to the surface.

Neo-Ddarwinsm. Darwin's theory of evolution through natural selection.

Neogastropoda. Order of Prosobranchia (Gastropoda: Mollusca); respiration by means of etenida; nervous system concentrated; sexes separate - most highly developed snails; species of *Columbella,* Conus, *Murex, Terebra* etc.

Neognathae. Superorder of Neornithes, an avian subclass; flying bii-ds having fully developed wings; no teeth; sternum with keel and fused caudal vertebrae.

Negregarinida. Order of Gregarinia (Sporozoa: Protozoa); unicellular minute organisms mostly insect-parasites.

Neoparatype. Paratype described subsequent to publication of descIiption of original type.

Neopseustida. Family of Micropterygoidea (Jugatae: Lepidoptera);

tiny moths having reduced mandibles without any teeth.

Neoteny. Phenomenon/wherein sexually mature larval stage becomes capable of reproduction without undergoing metamorphosis - some tailed amphibians like *Ambystoma* and salamanders.

Neotype. Specimen selected subsequently to serve as type specimen when all original specimens are missing or have been destroyed.

Nephromixium. Reproductive-excretory structure found in certain polychretes.

Nephros. Kidney: structure formed by several nephrons in kidney. that performs excretory function.

Nephrostome. Ciliated funnel-like opening at inner end of a nephridium through which excretory wastes enter the tubule.

Nephrotome. Mesodem of vertebrate embryo giving rise to nephron (kidney tubule).

Neritic. Oceanic sub-division of pelagic realm, the open water above continental shelf.

Nerve. Bundle ofaxons held together as a unit by connective tissue bundle of nerve fibres located outside central nervous system.

Nervous System. System formed by nerves, ganglia and brain etc., to act as co-ordinating system both for external as well as internal stimuli.

Neurentric Canal. Narrow connection between neural tube and archenteron of neurula in the postertor region.

Neurin. Filament found attached to the Inner surface of plasmalemma of nerve endings to the brain. It is similar to muscle action and may play a role in exocytosis (see neurostennin).

Neuroanatomy. Study of anatomy of nervous system and nerve tissue.

Neurocranium. Structure comprising of cartllagenous brain box and three sense capsules developed in connection with sensi' organs. This is developed in embryonic stage only.

Neuroendocrine Mechanism. Mechanism involving regulation of some body function through endocrine secretions which in turn are regulated by some part of the nervous system. As body blood sugar level is regulated by hormones secreted by islets of Langerhans. The secretion of which is in turn governed by PSH (pancreas stimulating hormone) from pituitary released under the influence of PSH-RF (pancreas stimulating hormone releasing factor) from hypothalamus.

Neurohumor. Synaptic transmitters and neurosecretory hormones.

Neurohyphophysis. Neurally derived reservoir for hormones with antidiuretic and oxytOCiC actions. Consists of neural stalk which is connected to and pass neurosecretions from hypothalamus and neural lobe which make up the bulk of neurohypophysis.

Neuromast. Collection of hair cells embedded in a cupula in lateral line mechanoreceptors of lower vertebrates.

Neuromuscular Junction. As axon (supplying skeletal muscle) approaches its termination looses its myelin sheath and divides into a number of terminal buttons which contain many small clear vesicles haVing acetylcholine. These endings fit into depression in motor end-plate. The muscle membrane and underneath is thrown into folds called palisades. The whole complex is called neuromuscular junction.

Neurons. Constituent cells of nervous tissue. Each has cell body/cyton or perikaryon containing nucleus and Nissl's granules *etc*~. and dendrites - that are one or more processes arising from cyton and axon which is a long process terminating at arrector or synaptic junction.

Neurophysins. Proteins associated with neurohypophysial hormones stored in granulesin neurosceretory terminals.

Neuropore. Closure of neural tube in neurula is not timed uniformly at all places. It starts from middle proceeding both anteriorly and posteriorly hence for sometime the neural tube remains open in front of the prospective brain box by the opening called neuropore.

Neuroptera. Order of ollgoneopteran insects (Hexapoda) with soft body. long antennoo. biting and chewing mouthparts. two pairs of long, similar, membranous wings having numerous longitudinal and cross veins; metamorphosis complete - ant-lions. alderflies etc.

Neurosecreton. Neurons which contain granules that strain with Gomori stain are generally called neurosecretory neurons and the secretion of chemical substance at their endings is called neurosection.

Neurostennin. Complex of two proteins found in brain nerve cell endings more possibly associated with exocytosis.

Neurotransmission. Passage of impulse from nerve to muscles. It is chemical at synapse. electrical in pre-synaptic region and at few of the conjoint synapses it is both chemical and electrical.

Neurotranmitters. Chemical mediators released by a post-synaptic nerve ending that acts with receptor molecules in ostsynsptic membrane. This induces permeability and influences electrical activity of postsynaptic cell *e.g.* acetylcholine.

Neutrophies. Type of leucocytes constituting 65 to 75% of total WBCs with granular cytoplasm and many lobed nucleus. Phagocytic in function.

Newt. Tailed amphibian. Triturus spp. (Urodela) that looks like lizard, legs well developed with four toes in forelegs and five in hind legs. no claws or webs; cloaca at base of tail;

terrestrial during non-breeding season 'aquatic during breeding seems -*T. verucosus* (East India and Myanmar>., *T. crtstatus* (Britain).

Nexus. Gap junction between cells characterised by 2 nm space between opposing membranes filled with densely packed particles through each of which there is a channel that connects each cell. it permits direct transfer of ions and other small molecules between cells with traversing the extracellular space.

Niacin. Member of vitamin-B complex group. Constituent of NAD and NADP. Found in yeast. meat. liver etc. Deficiency causes pellagra. It can be synthesised in body from tryptophan.

Niche. Constellation of environmental factors into which a species fits or which is required by a species.

Nictitating Membrane. Transport. thin membrane possessed by many vertebrates which opens and closes laterally across the cornea.

Nidamental. Gland in female cephalopod which secretes a protective egg capsule.

Nidicolous. Young ones of birds, helpless for some time and cannot walk and look after themselves soon after birth *e.g.*, chicken of sparrow: sharing nest of another species of bird.

Night-Apes. Small South African lemurs. *Galago* spp. and *Euoticus* spp. (Lorisidre: Lemurisiformes). arboreal and nocturnal mammals.

Night-Bird. Common name for any nocturnal bird especially owls and nightingales.

Night-crow. Indefined bird (Aves) of ill omen. which cries at night. NIGHT·DOG: dog that hunts during night.

Nightjar. Goatsucker : nocturnal migratory bird (Coprimulgidre: Aves) brown with white markings on long tall and wings;

lays eggs on ground; migrob~s to Britain during summer.

Nilgai. India blue bull; *Boselaphus tragocamelus* (Artiodactyla: Mammalia); male slaty-grey and female tawny.

Nipple Structure fonned by raising of skin at places and at which the ducts of mammary glands open. Made up of erectile tissue. Their number and position varies in different mammals.

Nissle's granules. Numerous bodies in the cytoplasm of cyton of a nerve cell. These are composed of ribosomes and RNA and produced under the influence of nucleus. They play important role in metabolism of nerve cell. They are absent in axon but present in dendrities particularly in the region of their origin.

NIT. Egg of body louse, found glued on to body hair of the host.

Nitrogen Narcosis. If a diver breathes ocmpressed air, the increased PN2 can cause symptoms resembling alcohol intoxication at greater pressures in which manual dexterity Is maintained but intellectual functions are impaired.

Nociceptive Stimuli. Flexor responses can be produced by Innocerous stimulation of skin or by stretch of muscles but strong flexor responses with withdrawal are initiated only by stimuli that are noxious or at least potentially harmful to animal, these are therefore called nociceptive stimuli.

Nocturia. Renal disease in which urine volume is increased resulting in waking up at night to void.

Noddy. Tropical sea-bird, *Anous Stoltdus* (Ares) brown or white in colour characterised by unforked taU; found on remote islands, feed on small fish but does not dive.

Node of Ranvier. Axon of nerve cellts constricted at places to form this structure. The portion of axon between two such nodes is called internode.

Nomeclature. Naming of organisms; binomial nomenclature

was first suggested by Carolus A Linnaeus (1707-78) and Is still being used.

Nomen Duplum. Taxonomic name that is invalid as It is not. accompanied by its description mudum proposed taxonomic name but invalid because its description given is not clear.

Non-Assortative Mating. Mating where mate selection Is random.

Non-Chordata. Animal lacking notochord in entire-life cycle *e.g.,* earthworm, cockroach etc. Compare with Chordata.

Non-Disjunctioin. Failure of chromosomes, chromatids to separate at the mitotic or meiotic anaphase. Non-disjunction Is responsible for several human chromosome anamolles.

Nonipara. Any woman who has undergone nine deliveries giving birth to alive or dead childern.

Non-Sense Mutation. Mutation which results in change from a codon specific for an amino-acid to a terminating chain -terminating codon which will bring translocation of a premature halt.

Nor-Adrenaline. Catecholamine secreted by adrenal medulla and also liberated at nerve endings. It serves to increase blood pressure by vasoconstriction, free fatty acid level, CNS stimulation, heat production and peripheral resistance and decreases cardiac out put.

Normoblast. Pre-erythroblast when becomes rounded and has Intense synthetic activity then it Is called normoblast.

Nose. Projecting part from centre of face; used for breathing, smelling and sometimes for producing sound.

Nostrils. Nares: one of the paired external openings of nose.

Nucleprotein. Molecular complex formed by nucleic acids and proteins.

Nucleosidase. Enzyme acting on nucleosides breaking union between purine or pyrimidine combined with pentose sugar (ribose or deoxytribose).

Nucleoside. Combination of purine or pyrimidine with pentose sugar (ribose or deoxyribose).

Nucleotidase. Enzyme acting on nucleotides (purine or pyrimidine combined with ribose or deoxyribose sugar and phosphoric acid) releasing phosphoric acid and resulting in the formation of nucleosides.

Nucleotide. A molecule that consists of phosphate, pentose sugar and purine or pyrimidine. *(see DNA).*

Nucleus. Cell organelle in eukaryotic cells which contains genetic material in the cell and is separated from cytoplasm by a nuclear membane. It controls all the activities of cell including reproduction. It divides amitotically, mitotically or meiotically. During cell division the nucleus undergoes division and one of the products (daughter nuclei) passes to each new cell. *Of Pander:* in the centre of oocyte there is a mass of white yolk which continues vertically as a neck-like structure and the neck is expanded into a disc-like area '".lying just beneath the blastodisc in the egg of a bird. This is called nucleus of pander. *Pulposus*: in mammals the mesenchyme cells between vertebral centra give rise to discs obliterating the notochord, however, it persists as nucleus pulposus or pulpy nucleus of intervertebral discs.

Nudicaudate. Having hairless tail - rat etc. NUDITY: state of being nude.

Nullipara. Woman who have had sex but not yet given birth to any child dead or alive.

Numbat. Banded ant-eater (Marsupialia : Mammalia), having brown fur with white bands, prominent ears an bushy tail; found in South-west Australia; feeding on ants and termites.

Numerical Taxonomy. Classification based on numerical analysis of variation of large number of characters in group of organisms assuming greater predictlvity with more characters used.

Nurse-Shark. Dogfish: large grey sluggish shark (Selachii : Pisces) often attacks human beings.

Nuthatch. Small song bird(s), *Sitta* spp. (Aves) having strong feet and claws creeps up and down tree trunk; feeds mainly on insects but in winter eats nuts, break opening the same with its sharp beak.

Nutrition. Sum of the processes concerned with growth, maintenance and repair of living body as a whole or its constituent parts.

Nyata. Large ox-like antelope *Tragelaphus angasi* (Artiodactyla: Mammalia) confmed to South Africa.

Nyctagmus. Rapid involuntary oscillating of eyeball.

Nyctalopia. Vitaminosis of vitamin A produces disturbances in synthesis of rhodopsin and iodopsin leading to night blindness.

Nyetatin. Rod-shaped antibiotic molecule that creates channels through membranes that allow passage of molecules of diameter less than 4A

Nympha. Immature young one of heterometabolic insect; similar to adult except smaller in size with wings and sex-organs not yet fully developed.

O

Oarfish. Thin compressed marine fish, that grows upto seven metres, bright red dorsal fin runs entire length of body forming conspicuous crest over the head; pelvic fins very long and slender with broad tips that look and work like oars.

Obesity. Pathological condition in human beings characterised by excessive deposition of fat in the subdermis and much gain in body weight and dimensions. It is due to disordered fat metabolism and usually starts with hyperphagia. This may ultimately lead to disorders of cardiovascular and respiratory systems as well as diabetes etc.

Objective. Lens or combination of lenses of microscope near to the object under observation.

Oblique Septa. A pair of tough elastic septa which separate thoracic cavities from abdominal cavity In case of birds.

Obliteration. Complete removal of organ or part of body by disease or surgical operation.

Observation. Perception, either qualitative or quantitative, of a set of phenomenon (the set may have only one member).

Obsolete. No more in usage; part of body that has become indistinct or imperfect and is no more in us.

Obturator Foremen. Foramen between ischium and pubis in the pervic girdle.

Occipital Arches. Pair of cartilagenous structures arising from parachordals behind the otic capsule and become attached to hind part of otic capsules during development of skull.

Condyle. Articulating surface in the back or posterior part of skull with the help of which first vertebra of vertebral column articulates. This is single in reptiles and birds but double In amphibians and mammals.

Odontoblasts. Mesenchyme cells in the pulp of tooth which secrete dentine. They have protoplasmic processes (Tome's fibres) which project through canaliculi of dentine tubles and deliver material necessary for maintenance of dentine.

Odontognathae. Extinet superorder of Neornithes (Aves) that included all large toothed, aquatic birds included in two orders -Hesperornithiformes and lchthyornithiformes, which were more advanced than *Archaeopteryx* but flightless, and good divers.

Oil-beetles. Beetles (Coeoptera: Hexapoda) that secrete yellowish oily liquid from its leg, when distured.

Oil-Bird. Guacharo: large brown bird spotted with white, nocturnal, liVing in caves in northern parts of South America: body contains large amount of fat, used as food by the natives.

Okapi. Rare, giraffe-like mammal (Artiodactyla), found in Congo forests of South Africa.

Olfactory. Sense organ showing capability of smelling objects. *Capsules*: cartllagenous structures in the skull of cartilagenous fishes like *Scoliodon* enclosing olfactory lobes and open on outer and inner lower sides and present in antilateral part just anterior to orbits. *Epithelium Schneiderian Membrane.* sensory epithelium forming olfactory receptor organ. Olfactory receptor cells are specialised neuro-secretory cells (bipolar neurons). These cells are spindle shaped and their cytoplasm contains Nissl's granules. Outer end of cell is long, thin olfactory rod bearing bunch of 5-12 delicate olfactory hair at tip and from other end arises axonic fibre and all such fibres constitute olfactory nerve. *Gland*: Bowmann's glands:

olfactory epithelium contains many mucous producing structures called olfactoly glands. *Lobes*: anteroventral parts of fore brain having power of smell (constitute rhinencephalon). *Nerve*: cranial nerve supplying ethomoid region of skull (nasal region) and arises from olfactory epithelium and terminating in the olfactory lobe to brain. *Epithelium Tract*: posterior stout portion of olfactory lobe leading to a slightly rounded elevation called tuberculum olfactorium.

Olfactory Ventricles. Rhinocoels: extension of paired cavities of cerebral hemispheres into olfactory lobes.

Ooligodendrocytes. Myelin containing neurons within the central nervous system.

Oligodendroglia. Myelin secreting cells of glia or neuroglia.

Oligolecithal Egg. Egg found in protochordates and mammals which is small (0.1-0.2 mm) with low concentration of yolk. The york is also dispresed.

Oligospermia. Sperm count below 20 million/ml in man.

Oligouria. Fall in urine volume due to destruction of most nephrons.

Olivocerebeuar Tract. Afferent tract in cerebellum that transmits proprioreceptor inputs from whole body via relay in inferior olive. Distributes all cerebellar coretex and deep nuclei.

Omasum. One of the divisions of ruminant stomach present between reticulum and abomasum.

Omentum. Structure (remnant of ventral mesentery) by which the alimentary canal in some regions is jointed to an organ.

Omnivorous. Eating all kinds of food, both animals and plants.

Omosternum. Inverted *'Y'* shaped bony structure attached to clavicles tipped by cartilage called episternum.

Onagadori Cocks. Tamed (domesticated) fowl of Japan with body hardly half a metre and beautiful flOWing tail upto 14 metres in length.

Onagri. Wild ass *Equus onager* (Perissodactyla: Mammalia); having broad brown stripe on the back; commonly found in Central Asia.

Onchosphere. Protective layer around hexacanth embryo of tape-worm *Taenia solium* (Cestoda: Platihelminthes).

Onchogenic Virus. Virus which can induce cell proliferation and tumour formation.

Oncotic Pressure. Colloid osmotic pressure due to plasm colloids. ONTOGENY: entire developmental history of the individual organism.

Oocyst. Swelling in stomach wall caused by implantation of sporozoites.

Opcyte. Egg mother cell from which are produced secondary oocyte and first polar body by first maturation division.

Oogenesis. Process of ova formation.

Oologist. Person having good knowledge of oology.

Oology. Collection and study of avian eggs, a branch of zoology.

Ootype. Thickening of oviduct near ovaries in liver flukes and tapeworms.

Oozoid. Tailed larva developing from ovum of a sexual zooid in some protochordates like *Doliolurn.*

Opah. Kingfish; moonfish (Lamprididre: Pisces), deep-bodied sea fish. body laterally compressed covered with minute scales and having paired sickle-shaped fins; grows upto 1.4 metre in diameters; native to West Africa.

Open System. System which is not self-sufficient. Most biological systems are open since they are energy requiring (take energy from environment).

Operator Gene. Gene that regulates synthetic activity of closely linked structural genes via its association with a regulator gene.

Operculum. Structure covering gills of bony fishes and tadpole; also platelike structure serving to cover the opening of some snail shell and the lid-like structure present in the helminthes eggs.

Opercular Chamber. Spaceous cavity in bony fishes *e.g.,* Labeo lying between gill slits and operculum communicating to exterior through a single aperture. *Plates*: four plate-like structures supporting opercular flap. These are opercular, preopercular, subopercular and interopercular plates.

Operon. Unit of DNA consisting of operator site and one or more structural genes whose activity is influenced by the operator.

Ophydiarium (*pl. Ophidiaria*). Snakehouse, place where snakes are kept for exhibition and rearing.

Ophidia. Suborder of Squamata (Reptilia) to which snakes belong.

Ophthalmic. Anything pertaining to eye like opthalmic artery. ophthalmic nerve. *Nerve*: sensory nerve and one of the branches of 5th cranial nerve (trigeminal) supplying skin of upper eyelid, skin of front of head and inside as well as outside of nose.

Opinicus. Mythological monster. half lion and half dragon.

Opossum. Small tree dwelling *marsupialDtdelphys* species (Mammalia) of America; 'tail as long as body. each 30 to 50 cm long; marsupium opening ventro-posteriorly

Opisthonephros. Kidney in some fishes like *&oliod.on* is basically mesonephros but the tubules in adult shift towards the posterior side of body giving kidney a shape that looks similar to metancephros. In such a kidney anterior portion is non-renal and reduced while the posterior portion is functional and contains renal tubules.

Opsin. Protein forming photosensitive compound in human eye and other mammals.

Opsonization. Process of making bacteria tasty (in plasma) to phagocytes. *Axis:* Imaginary straight line passing through the centre of curvature of simple lens. *Capsules*: paired cartilagenous structures formed on each side of cranium during development of skull. Unlike olfactory capsules these do not fuse with brain box and remain cartilagenous in many vertebrates. Free optic capsule form the sclerotic coat of the eyeball. *Chiasma*: X-shaped structure formed by crossing of optic nerves in the mid brain. Part of optic tract where neuron from medical retinae decussate. while those of lateral retinae merely continue ispsilaterally. *Lobes*: thickenings on the dorsal surface of brain. Usually two (forming corpora bigeminal but in mammals they are four in number (forming corpora quadrigeminal. *Nerves*: nerves from eye to the brain. *Stalk*: narrow passage formed by constriction of proximal portion of optic vesicle. It serves to attach the eyeball.

Optocoel. Cavity/ventricle/space in the optic lobes or mid brain. *Hood*: in animals like *Amphioxus* ventral side of rostrum is expanded into a hood-like structure. which is called oral hood. *Groove*: groove having ciliated ridge starts at anterior one-third of ventral part of body and runs upto almost middle in some animals like *Paramecium* species. *Valves*: pair of folds in mouth of certain bony fishes like *Labeo* which arise from a maxillary and mandibular region and permit water entry in buccal cavity but do not allow back-flow.

Oraslerrata. In retina of eye light sensitive part is demarcated from the non-sensory part by an irregular and serrated area called ora serrata terminals.

Orbit. Eye pit or eye socket: cavity in frontal portion of skull that contains the eyeball.

Orb-Spider. Diadem-spider : common name for such spiders (Arachnida: Arthropoda) that spin orb-shaped webs wherein they catch their prey (insect). paralyse it with their poisonous fangs and wrap it in silken threads before feeding on the same - garden-spider.

Order. Category in taxonomic hierarchy comprising of one or more related families.

Organ. Multicellular part of animal formed by tissue and performing a definite function: group of tissues associated with performing one or more functions in the body - kidney.

Organ-pine Coral. Aleyonarium coral (Coelenterata) with long paralleled skeletal tubes composed of fused spicules.

Organelle's. Specified part within single cell differentiated to perform certain functions and comparable to an organ as found in the case of bigger animals. These structures can divide and take part in cellular activities. Their presence is a must for the cell.

Organiser. Part of the embryo which influences some other part and directs its histological and morphological differentiation.

Organium Vasculosm of Lamina Terminalis. When an acidic dye is injected into an animal a small area among five areas in brain stain like tissues outside the brain. Also called supra optic crest and forms part of circumventricular organ.

Oribi. Pale-buck: small (dwarf) ontelope (Artiodactyla: Mammalia) 60 cm shoulder Height; having straight horns; found in grasslands of Africa. south of Sahara.

Oriole. Group of brightly coloured song-birds of Old World *Oriaous* spp. (Oriolidce Aves) : Unrelated American birds of lcteridace family having yellow or orange and black plumage.

Ornis. Avifauna: birds of a region.

Ornithine Cycle. Urea cycle: a cyclic succession of reactions that f:;liminate ammonia and produce urea in the liver of ureotic animals like.frog and eutherian mam!nals such as man.

Ornithologist. Person having sound knowledge of ornithology.

Ornithology. Branch of zoology that deals with study of birds only.

Ornithomancy. Divination by observing flight of birds.

Ornithosaur. Extinct bird-like dinosaur: that lived in Juraosic and Cretaceous periods.

Ornithoscopy. Anguxy by observation of birds.

Orthogenesis. Evolution of organisms systematically in definite direction and not accidentally in many direction.

Orthoptera. Order of insects (Hexapoda) having biting mouthparts. foreWings forming thickned tegmina, hind legs enlarged and adapted for jumping; auditoxy and stridulatoxy organs usually present. metamorphosisgradual - crickets. grasshopers. locusts etc.

Os-Coruna. Two bony buttons in newly born calf which can be felt under skin just opposite to frontal bone.

Os-Innominate. Each half of pelvic girdle in animals like *Varanus* or rabbit.

Os-Penis. Small bone present In the septum between corpora cavemosa in many mammals; it provides additional rigidity to the penis.

Osmoconformer. Organism that exhibits little or no osmoregulation so that the osmolarity of its body fluids follows change In osmolarity of environment.

Osmometer. Instrument for measuring osmotic pressure.

Osmoregulation. Maintenance of internal osmolarity with respect to environment by an animal. Balance of water and salts within the body of an animal.

Osmorgulator. Organism that controls its internal osmolarity in event of changes in the environmental osmolarity.

Osmosis. Movement of solvent molecules across a semipermeable membrane. When two solutions of different concentrations are separated by such a membrane or a solution is separated from solvent by semipermeable membrane.

Osmotic Potential. Depression in chemical potential of a solvent due to presence of solute molecules in it. It has always negative value.

Osmotic Pressure. Hydrostatic pressure necessary to check the movement of solvent molecules across a semipermeable membrane in a system where osmosis is occurring. It always has a positive value.

Osphradium. Sensory area in incurrent siphon in some molluscs, epithelial and frequently yellow.

Osprey. Fish-hawk: large diurnal bird of prey Pandion *haliactus* (Falconiformes: Aves). nests in trees near sea- shore, large lakes and rivers..and feeds on fish.

Ossein. Protein found in the matrix of bone.

Osteichthyes. (Actinoptreygu) class of pisces including bony fishes *Lata.* Labeo. rahu, sea-horse. trout etc.)

Osteoblasts. Cells that secrete fibrous matrix of the bone.

Osteoclasts. Giant cells formed by fusion of several osteoblasts which begin to eat away bone and remove the unwanted spongy part of disphysis which becomes replaced by bone marrow.

Osteologist. Person having sound knowledge of bones especially the formation. organisation and skeletal system: bone-specialist.

Osteology. Scientific study of bones.

Osteomalacia. Disorders like rickets in which amount of mineral secretion in matrix of bene is deficient leading to deformations in bones.

Osteon. In mammals large bones have a system formed by a single Haversian canal and surrounding lamellae containing osteocytes.

Osteoporosis. Decrease in bone mass with preservation of normal ratio of minerals to the matrix.

Osteoscellerosis. Presence of increased amount of calcified bone occurring in matastatic tumours. leads to poisoning and hypoparathyroidism.

Osteoscutes. Large dermal scales found in turtles and tortoises where they form a box-like shell. Often they are of small size and unfused and found over the throat of crocodilians. Lizards and snakes also have these under epidermal scales.

Ostium. Small mouth-like opening such as anterior end of oviduct in frog and other vertebrates; in general an opening in both invertebrates and vertebrates usually guarded with a valve or circular muscles.

Ostracoda. Subclass of Crustacea comprising of minute bivalved aquatic creatures having unsegmented body without any true abdomen - *Cypris* species.

Outer Segment. Portion of a vertebrate visual receptor that contains pigmented receptor membranes. It is attached to inner segment by a thin cilium-like connection.

Outer Segment-Oval. Broadly elliptical, round at both ends, nearly twice as long as broad and broadest in middle.

Ovate. Egg shaped; broadest below the middle portion. OVIBOS: musk-ox.

Ovibovine. Animal having characteristics between sheep and ox. OVICELL: broad chamber in certain bryozoans.

Ovid. Any species of animal oelonging to family Ovidce (Mammalia). OVADLE: family of ruminant mammals, quadrupeds having unbranched horns covered with thick keratin - goat, sheep etc.

Oviduct. Tube that conveys eggs from ovary to uterus or to outsine.

Ovipositor. Organ in female insects that helps in deposition of eggs. OVISAC: egg-capsule: brood pouch.

Ovoviviparous. Producing eggs that hatch within the parent's body but are not nourished by mother through placenta marsupials.

Oovum. Non-motile female gamete, especially of insects, fish and mammals.

Owl. Nocturnal predaceous bird (Stringiformes: Aves) large broad head, flat face, large eyes surrounded by discs of feathers, short hooked bill, howling or hooting cry.

Owl-Monkey. Douroucouli only nocturnal monkey with large eyes and owl-like face: found in rain forests of South America feeding of small animals.

Ox. Domestic bovine animal. *Ox-anielope*: any antelope of hartebeet group. *Ox-bird*: ox-peeker: an Mrican weaver-bird (Aves). *Ox-bot*: warble-fly (Hexapoda) larva infesting cattle. *Ox-Eye*: small common bird.

Oxpecker. Bird often seen perching on back of large Mricart mammals. feeding on flies. Fleas and ticks of its host. thus benefiting its host by removing its parasities.

Oxidative Phosphorylation. Conversion of inorganic phosphate to energy rich phosphate of ATP by reactions coupled to transfer of electrons in electron transport system of mitochonodria.

Oxygen Debt. Accumulation of lactic acid in muscles during exercise.

Dissociatioin Curves. Curves that describe the relationship between the extent of combination of oxygen with the respiratory pigment and partial pressure of oxygen in gas phase.

Oxyhaemoglobin. Unstable chemical compound in blood formed by combination of oxygen with haemoglobin. When it goes to tissue the oxygen is released due to dissociation of the compound.

P

Pacca. Spotted rodent, *Cuniculus paca* (Rodentia: Mammalia) found in Central and South America; lives in burrows near water and feeds on fallen fruits and leaves.

Pace Maker. Principal area in the heart from where there is discharge leading to depolarisation which ultimately brings contraction of heart. In mammals SA-node (sinuauricular node) is said to be principal pace maker.

Pacyglossal. Having thick tongue (Lizards). *Pacinian Corpuscles*: heavy touch receptors. *Paddle*: fin or ilipper.

Paddy-bird. Rica-bird: a small sparrow. *Passer* Species (Passeriformes: Aves) found in Java.

Paedogenesis. Development of sexual maturity and reproductive capacity of larval or other pre-adult stages; found 'in some insect species.

Paedologist. Person having sound knowledge of paedology. *Paedology* : scientific study of physiological and psychological aspects of childhood.

Pa-Interval. Unit in electrocardiogram indicating depolarisation and conduction through AV node (auriculo ventricular node). Normal time (duration) being 0.18 seconds.

Palaenthrous. Extinct human beings including Neanderthal and Heidelberg races.

Palae-ethnologist. Scholar of palae -athnology.

Palae-ethnology. Sdentific study of early and extinct human races and their evolution.

Palaeichthyologist. Person well-versed in palaeichthyology.

Palaeoconche. Order of extinct. smooth-hinged bivalve molluses.

Palaeoecology. Branch of life sciences dealing with prehistoric ecology as revealed by study of fossil.

Palaeoherpetalogist. Zoologist who has specialised in palaeoherpetalogy.

Palaeoherpetalogy. Study of fossil reptiles.

Palaeonisciformes. Order of extinct fishes including earliest known primitive ray-finned fishes.

Palaeontologist. Person having sound knowledge of palaeontology.

Palaeontologist. Person specialised in palaeontology.

Pandaka. Pigmy smallest fish (Pisces) in the World not bigger than grain of rice; found near Phuippine Islands.

Paneth Cells. Among cells of Krypts of Lieberkuhn there are some argentaffin cells which probably produce enzymes.

Pangolis. Scaly ant-eaters; large burrowing animals *Manis* spp. (Pholidota: Mammalia) having snout elongated, eyes small, ears vestigial, teeth absent, tongue thin long and glutinous, lower body surface covered with longitudinal rows of horny epidermal scales: feed on ants and termites -M. *crassicandata* is found in hilly tracts south of Himalayas while M. *Qurita* is confined to Assam. Sikkim and Nepal.

Panthothenic Acid. Member of vitamin-B complex group and constituent of coenzyme-A. It is found in eggs. liver. tomatoes etc. and its deficiency causes dermatitis. enteritis (skin and alimentary canal disorders).

Papilionid. Samily of Papilionoidea (Lepidoptera) large conspicuously coloured butterflies with tail-like prolongation of hind wings; caterpillars voracious feeders on leaves of various crops. *Papilto* spp: are most destructive pests of citrus trees.

Papillae. Structurt.; in the tongue on which taste buds are located. Several types- fungiform papillae which are rounded and most numerous near the tip of tongue. vallate papillae are found in *V* on the back of tongue and filiform papillac covering the dorsum of tongue. Taste buds are 5 per fungiform papillae. The large vallate papillae may have upto 100 taste buds while filiform papillae usually do not contain taste buds.

Parabiosis. Experimental connection of two individuals to allow mixing of their Lody fluids.

Parabronchus. Branches of ecto-and endobronchi in respiratory system of birds which enclose air capillaries and where the exchange of gases takes place.

Paracanthopterygii Super order of Teleostei (Pisces) that includes higher bony fishes like cod etc.

Parachordals. Pair of flattened and curved cartilages differentiated from mesenchyma on each side of notochord and beneath the brain during development of skull.

Paracoels. Paired cavities of cerebral hemispheres. Also called lateral ventricles.

Paracondyloid. Process on outerside of each condyle of occlpital bone in skull of certain mammals.

Paracrinoidea. Class of fossil echinoderms (Crinozoa) that had numerous irregularly arranged plates and uniserial arm-Uike appendage.

Paradoxical Sleep. Sleep state wherein rapid eye movements occur (REM-sleep); Electroence Phalogram tracings look as if the individual is awake.

Paratype. Specimen other than holotype or isotype available with author at the time of describing the said species as new species.

Paraxonic Foot. Foot in which third and fourth digits are dominant and carry most of body weight while 2nd and 5th digits are reduced or vestigial and displaced posteriorly and the first digit is not there *e.g.*• pig. Compare with mesaxonic foot.

Parenchayma. Loose. spongy. connective tissue in invertebrates. In vertebrates, the specific tissue component of an organ such as liver (hepatic cells) etc.

Parkinson's Disease. The syndrome originally described by James Parkinson in which there are pathological changes in nigrostraital system of dopaminergic neurons. Characterised by poverty of movement and hyperkinetic features like rigidity and tremor, difficulty in initiating voluntary movement and decrease In normal unconscious movements like swinging of ann during walking and expression related to emotional content of thought and speech etc.

Parotid Gland. Set of salivary glands in the region of ear just beneath external auditory maetus. Its duct is called Stenson's ductinman.

Parrakeet. Small brightly coloured parrot. *Pstttacula* spp. (Psittaciformes: Aves) having articulated beak, long tail and syndactylus feet adapted for climbing branches of trees; feed on ripening fruits.

Parrots. Common name of over 300 species of birds including cockatoos (Australia), lorikeets (Malaysia). love-birds. (Africa). macaws (tropical America) and Parackeets; all are stocky birds .with large head. strongly hooked beak; feed mainly on ripening fruits and seeds.

Paras Basilaris. Outgrowth/outpocketing of lagena in membranous labyrinth of certain frogs in which there may be basilar papilla (macula).

Pedicellaria. Small pincer or. scissor-like processes on the surface of certain echinoderms like sea-stars. sea-urchins

Pedicle. Narrow waist between cephalothorax and abdomen in spiders.

Pedipalp. Either of second pair 01 appendages in arachnida, often sensory and sometimes used in seizing prey as in scorpions.

Pedipalpus. One of second pair of appendages of arachnid. *Pedogenesis.*: reproduction by larvae. Also called neoteny. Found in axolotl larva of certain amphibia like *Ambystoma* spp.

Peewit. Lapwing or green plover. *Vanellus rxznellus.* a small wading bird with a crest on its head.

Peke. Pekingese dog. *Canis* species (Carnivora: Mammalia) dwarf pug-dog having long silky coat. flat face and prominent eyes:

Peking Man. Extinct human race *Sinanthropus pekinensis (Homo & erectus stnanthropus)* that. had thick brain box and heavy browridges.

Pelecanidat. Family of Pelecaniformes (Aves) - Pelicans .. *Pelecaniformes*: order of aquatic fish-eating birds (Carinate: Aves) adapted for diving and fishing: characterised by having all four toes joined together-cormorants. pelicans, spoonbUls etc.

Pelecypoda. Class of Phylum Mollusca that have hatchet-like foot and bivalve shell, pearls are secreted by the mantle of some fresh-water mussels and oysters and these lie within the body between mantle and shell, used as food delicacy and in manufacture of pearl buttons novelties and jewellery - calms. mussels. oysters etc.

Pellagra. Disease caused by deficiency of niacin in body. It is characterised by dermatitis, diarrhoea and dementia (3 0-disease).

Pellet. Mass of undigestible material regugited by carnivorous birds like hawks. owls, etc.

Pellicle. Protective layer on the surface of some protozoa like Paramecium, *Monocystis* etc.

Pelmatozoa. Subphylum or division of Echinodermata that includes such small marine animals as are anchored to substrate during entire or part of their life (both extant and extinct forms).

Pelodytida. Family of Anomocoela (Anura: Amphibia) that includes frogs.

Pelvic. Group of bones to which bones from hind limbs are attached.

Pelvi.-Girdle. System of bones supporting hind or lower limbs of vertebrate animals. *Veins*: veins branching off from femoral veins in amphibia and some reptiles. carry blood forward to liver.

Penetrance. Ability of a genotype to be expressed in any way at all when present. If a gene or genotype is penetrant that expression may vary (be of variable expressivity).

Penis. Male copulatory organ used in coition for transferring sperms into female genital tract.

Pentastomid. Uncertain arthoropoda. soft worm-like animals with short cephalothorax and elongated ringed abdomen; sexes separate; endoparastic. in vertebrates-Sibekia in crocodiles; *Linguatula* serrata In goat, horse. rabbit etc.

Pennae. Feathers visible over the body of a bird. *Pentadactyl*: having five fingers, toes or digits.

Pepsin. Enzyme concerned with protein digestion in animals. *Pepsinogen*: inactive form of pepsin in which it is secreted in order to prevent autodigestion of body in individual. It is activated by acidity of food in stomach.

Peptide. Bond formed between two amino-acids when these join end to end resulting in depeptide. A polypeptide results from joining of many amino-acids and is primary component of a protein molecule.

Perch. Edible. spiny-finned fresh-water fish *Petcajluuiatilis* having dark vertical stripes; commonly found in Europe.

Percopsiformes. Small order of Actinopterygtii (Osteichthyes: Pisces) with single, ray supported dorsal and anal fins and pelvic fin with three to eight soft rays.

Perfusion. Passage of fluid over or through an organ, tissue or cell.

Periblast. Layer of cells under blastocoel infishes. *Pericardio-peritonial Canal*: perforation in the transverse septum which separate abdominal cavity with pericardium in fishes.

Pericardium. Closed membranous area surrounding the heart. The part in contact with the heart is visceral pericardium and the other is pareital pericardium.

Peri Centric Inversion. Inversion involving both arms of a chromosome, the breaks being on the opposite side of centromere as a result, centrometre's position can be shifted.

Perichondrium. Fibrons connective tissue layer covering the surface of cartilage.

Peroneus Longus. Superficial flexor muscle in the leg of perching birds.

Pessulus. Cartilage like structure in the voice box of birds (syrinx) over which a semilunar membrane is mounted.

Fest. An organism (animal or plant) that causes damage. loss or annoyance to man or his property to the extent that wasrants its control.

Pesticide. Chemical used for killing the pest; it may be acaricide. bactericide fungicide. herbicide. Insecticide. molluscide. nematocide. rodenticide. viricide. Weedicide etc.

Petrohayals. Set of five bands of muscles arising from outer surface of otic capsule and pass in fan-like manner around floor of phaynx and oesophagus and inserted into the side of hyoid. They raise floor of mouth cavity during pulmonary respiration in frog.

Petromyzontia. Order of Cyclostomata (Agnatha) having long cylindrical body with suctorial funnel at anterior end bearing horny teeth and raspring tongue; larva microphagous. adult parastic - species of *Lampetra. Mordacta Petromyzon* etc.

Phenocopies. Different phenotypes produced from one genotype by manipulation of environment *e.g.,* in Himalayan \rabbit the phenotype reared at 30'C becomes colourless and if reared at a temperature below 10'C their fur becomes black but if part of skin is shaved and cooled artificially below 25'C by continued use of ice packs, the new fur will be back in limited area.

Phenology. Study of organisms (plants and animals) as affected by climate - dates of seasonal arrival and departure of migrants.

Phenocopy. Environmentally induced non-genetic initiation of the effects of specific genotype.

Phenotype. Expressed trait of an individual as contrasted with genotype. The total sum of the structural and physiological traits. Sometimes incorrectly defined as external appearance of individual which would exclude physiological traits such as blood groups. Compare with genotype.

Phenylketoneuria. Recessively Inherited condition characterised by the inability of organism to convert phenylalanine to tyrosine resulting in accumulation of toxic products which may cause mental retardation unless dietary precautions are instituted at proper time.

Pheromone. Species specific substance released into environment for purpose of signalling between individuals of the same species.

Phonoreceptor. Sense organ responsive to sound.

Phorozooids. Median buds from cadophore of certain protochordates like *Doliolum* that develop muscle bands and propel entire community of buds.

Phosphagens. Compounds that store light energy phosphates like phosphocreatine.

Phosphatases. Hydrolytic enzymes that act on acid and alkaline phosphatases and release phosphates from such compounds.

Pine Grosbeak. Finch-like bird (Passeriformes: Aves), with large conical beak.

Pineal Apparatus. Behind the anterior choroid plexus the roof of diencephalon gives off two median projections, anterior parietal body and posterior pineal body. The two collectively constitute pineal apparatus or epiphyseal apparatus. *Body*: structure on the roof of brain of vertebrates. It is phylogenetic ally associated with medium eye.

Pinion. Distal portion of bird's wing: any flight-feather of wing. *Pinna*: external ear in mammals.

Pinnipedia. Suoorder of Carnivora (Mammalia) that includes paddle-footed aquatic animals - otters, sea-lions seal and walruses.

Pinnotheridae. Family of Brachyura (Decapoda: Crustacea) - pea crabs etc.

Pomtao. Surface feeding migratory duck, *Anas* species (Anseri formesi Aves); drake having a pair of long beautiful tail feathers; found in northern hemisphere in summer but goes down south to pass winter.

Pinworm. Threadworm or namatode *Enterooius vemicularis* (Oxyuroidea: Nematoda): parasitic in human rectum).

Pipe-Fish. Long thin fish (Osteichthyes: Pisces) covered with hard plates; the jaws forming a long tube.

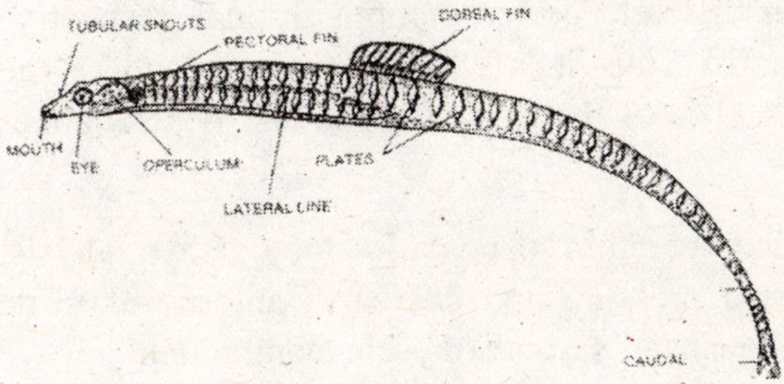

Pipit. Common name for group of small birds. *Anthus* spp. (Passeriformer: Aves) found in open country allover the world -meadow pipit. *A. pratensis.*

Pipridae. Family of Passeriformes (Aves). colourful. neotropical suboscrine birds - manakins.

Piranhha. Tropical. voracious fresh-water fish. *Surrasalmo piraya* (Pisces) having serrated belly and strong lancet-shaped teeth.

Pisces. Superclass of Chordata; poikilothennal vertebrates. breathing by means of gills (except Choanchthyes). streamlined body adopted to aquatic mode of life. skin usually covered with scales. pectoral and pelvic fins paired; and median fins unpaired no tympanic membrane - t1shes.

Pisciculture. Rearing fish by artificial methods. PISCIVOROUS: fetciing on fish.

Pittidae. Family to Tyranni. brightly coloured suboscine birds.

having erectile crown of feathers.

Pituicytes. Stellate cells of posterior pituitary lobe contaIning fat globules which are now regarded as modified astroglicll cells but considered previously to screte hormones.

Pituitary. Endocrine gland located on the ventral surface of brain and composed of anterior. intermediate and posterior lobes. It secretes a number of hormones which exrecise co.'l.trol over different body functions through respective endocrine glands. Thus it is called 'Master Gland of the Body'. In man it is of the size of a pea and weighs about 0.5 gIn.

Pituitrin. Combined secretion of posterior lobe of hypophysis, raises the blood pressure and decreases amount of urine formation and causes smooth muscle contraction.

Pit-viper. Group of poisonous snakes characterised by having deep bit on each side of face between eyes and nostrils; bite is very painful and often fatal.

Placenta. Organic bridge/connection derived from allantois and chorion in a eutherian mammal between body of mother and developing embryo acting as an organ for nutrition, respiration and excretion. It consists of two parts - maternal part which is derived from mother and foetal part which is derived from embryo.

Placodermi. Armoured fishes: large and varied class . of fossil fishes (Pisces: Chordata); oldest jaw-verteriates that appeared during end of Silurian period and disappeared in beginning of Devonian period. *Climatius* (climatiformes).

Placodont. Fossil reptile from Triassic period which possesses flat crushing teeth and feed on molluscs.

Placoid Scales. Scales found in cartilagenous fishes with basal plate and rearward pointing spine.

Plaice. Yellow-spotted. edible. marine flatfish *Pleuronectes platessa*

(Pleuronectidae: Pisces) of Europe. found in shallow water. resting on one side.

Pleiotropic. Pertaining to morphological effects of a single gene and dominant recessive relationship.

Pleitropy. Multiple phenotypic effects of a single allele.

Plesiosauria. Group of great Jurassic period fossil sea-reptiles (Sauropterygia), 3 to 15 metres long, with very long neck and short tail; four flippers, used as oars for swimming.

Pleura. Strong sheet along ventral surface of avian lung. Also called pulmonary aponeurosis. Also used to designate coelo'Ilic epithelium layers of the compartments enclosing lungs in mammals.

Pleural Cavities. Cavities between pleural membranes of mammals which form double layered jacket around each lung. These cavities contain in lymph-like fluid which prevents sticking of pleural membranes and allows the lungs to slip over the membranes during breathing.

Pleuron. Lateral part of cuticle in each segment of a crustacean or insect.

Pleurotremata. Group of certilagenous fishes with lateral gill slits. commonly called sharks. *e.g., Scoliodon* species.

Plumage. Natural covering of feathers, as in birds: feathers collectively.

Plumate. Feathered or adorned with feathers. *Plume*: large showy feathers: vane of feather.

Moths. Group of small moths (Pterophoridre: Lepidoptera); forewings are plume-like. feed on various economic crops -*Exelastis atmosa* and *Oxyptilus rugulus* are pests of grapevInes etc.

Polyembryony. Phenomenon of producing several larvae in the life-cycle as evidenced in the life-cycle of *Easciola.*

Polygenic Inheritance. Inheritance in which many genetic loci are involved which interact with environment. Polygenic refers to many (specially) genetic factors as opposed to environmental ones.

Polylecithal Egg. Egg with large amount of yolk as found in birds and egg laying mammals.

Polymer. Compound of linear sequence of simple molecules or residues.

Polymorpffic. Organism having several different forms/castes.

Polyneoptera. Insects having wings with large number of veins arranged in elaborate branching pattern and no cross-veins.

Polyphaga. Large suborder of Coleoptera (Hexapoda); beetles with head not prolonged into beak, prosternal structure distinct, and hind wings without cross-veins.

Polyphagous. One who eats different kinds of food.

Polyphagy. Ability to feed on wide range of unrelated hosts.

Polyphydont. Having teeth that are constantly replaced.

Polyp. Sessile coelenterate having hollow cylindrical body; mouth surrounded by tentacles at free end - hydra, coral etc.

POLYPLACOPHORA: order of molluscs; elliptical body with dorsal shell composed of eight calcareous plates - chitons.

Polyribosomes. Polysomes: m-RNA chain and its collection of ribosomes visible under electron microscope.

Polysaccharides.. Carbohydrate composed of numerous joinpd monosaccharide units - glycogen.

Polytene. Having many duplicate chromatin strands.

Polyuria. Passage of large amounts of dilute urine.

Population. Group of individuals sharing defined common

characteristics (geographical. genetic etc.).

Precavals. Symmetrical large veins draining head, neck and fore limbs and pouring into right auricle separately (Also called superior vena cava).

Precocious. Early in reaching some stage of development as mental maturity.

Predation. Biotic interaction involving association between a strong and a weak animal.

Predator. Individual that lives by preying upon other animal species - dragonflies, lizards etc.

Preen Gland. Branched tubular glandular structures in birds placed at the end of tail. Also called pyginial glands. The oily secretion of these glands is coated over the feathers by birds (by rubbing the secretion with the help of its beak) to make them waterproof.

Prefrontal. Bone of some vertebrate skulls situated anterior and lateral to frontal bone.

Pregnancy. Processes right from fertllisation to the maintenance of embryo upto birth in human beings are collectively called pregnancy.

Pregnant. Female having offspring developing in her uterus.

Pregneralon. One of the intermediates formed during the .biosynthesis of testesterone from chlesterol.

Prehensile. Structure adapted for holding/grasping some object such as prehensile tail in many animals like *Chamelion. Hippocampus,* langoor etc.

Prekallikrein Activators. In fibrinolytic system the substances involved in the formationof kinin and formed by the action of plasmin on Clctive factor-XII.

Premolar. Grinding teeth of mammals with broad, fiat crown and many roots. They are represented in milk dentition unlike molars.

Prepuce. Foreskin of penis: similar fold over glans clitoridis in female.

Presbyopia. Condition in human beings leading to difficulty in reading and close work by the age of 40-45 years owing to receding of near point of eye. It can be corrected by using convex glasses.

Prezygapophysis. Articular surface near the anterior phase vertebrae and directed upwards.

Priapulida. Minor phylum of Worm-like marine animals having proboscis, trunk and I caudal appendages; covered with spines and tubercles.

Priapuloidea. Old order of invertebrates, now raised to phylum Priapulida.

Primates. Order of mammals (Eutheria: Chordata) ; that includes animals having typically pentadactyl limbs for grasping, both eyes facing forward. small olfactory lobes, large cerebral hemispheres -and nails instead of claws; it embraces four suborders - Lemuriformes (lemurs), Lorisiformes (lorises), Trsiformes (tarriers) and Anthropoidea (monkeys, apes and men).

Primitiopascea. Superfamily of Beyrickiopina (Paleocopa). extinct ostracods that had velum of male narrow and uniform while that of female was greatly expanded posteriorly.

Primitive Knot. Prechordal and chordal cells form distinct swelling at the anterior end of primitive streak which is called primitive knot or Hensen's knot. Later it streams forward to form head process.

Primordial. First stage or first in order of time; the primitive form.

Primordium. Earliest stage of formation of a body organ or part.

Proboscis. HTubular extension of nose or lip or pharynx. Also extended break-like mouthpart in certain insects.

Procellariformes. Order of ocean-birds (Carinatre: Aves); distingUished by tube-like nostril openings. horny beak, dense plumage and webbed feet - albatrosses, petrels etc.

Procencephalon. One of the divisions of neural tube giving rise to olfactory lobe. cerebral hemispheres and diencephalon.

Proctodaeum. Terminal part of digestive tract near anus and lined with ecoderm.

Progesterone. Vertebrate female sex hormone secreted by corpus luteum and placenta (during pregnancy) responsible for maintenance of uterine cycle, maintenance of pregnancy and changes associated with reproductive cycle in female.

Progestins. Progestational agents: gestgens: substances m1m1c1ng the action of progesterone in female; when administered in the body orally in appropriate dose these block mid cycle burst of LH secretion and act as contraceptives.

Prolactin. Hormone from pituitary (also secreted by corpus luteum of pregnancy) which stimulates secretion of milk from mammary glands: hormone secretion is influenced by PRH from hypothalamus.

Prolegs. Small stumpy abdominal legs of caterpillar, disappear during pupal stage and are not found in adult insect.

Prometaphase. Stage during which there is disappearance of nuclear membrane and completion of spindle formation during mitosis / meiosis.

Promiscuous. One having sexual intercourse with many females of his species.

Pronate. To turn palm backward or downward with radius and ulna crossed.

Pronation. Placing of palm facing ground: act of pronating.

Prone. Condition of fixing hand during posture (palm facing the ground) : prostrate.

Pronephros. In primitive vertebrates the first kidney formed embryologically by tubunephrotomes of anterior part of embryo.

Prongbuck. American antelope-like ruminant. *Anttlocarpa* americandt'llaving deciduous horns pronged in front.

Pronghorn. Horn in antelope basically resembles the hollow horn. Two bonystumps of frontal bones covered with skin which secretes the horn. It is shed periodically.

Pronotum. Dorsal surface of an insect's pro thorax including--sclerites.

Protozoon. Microscopic organism consisting of single cell or colonies of like cells multiplying by fission.

Protura. Order of very. small, wingless insects (Hexapoda) mouthparts degenerated; eyes and antennre lacking; and metamorphosis slight or wanting - telsontat1s etc.

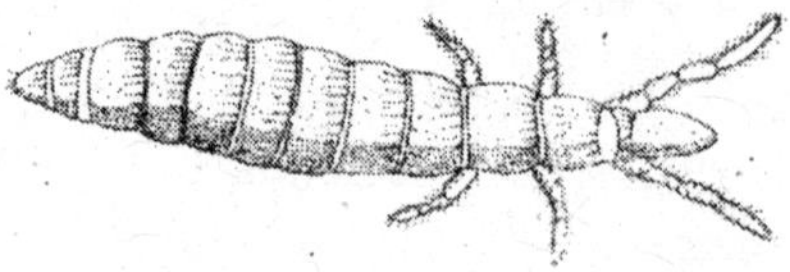

Proventiculus. Anterior portion of avian stomach. gizzard: anterior glandular part of stomach. often having strong cuticular plates or teeth on its inner walls.

Proximal. Nearest point of an organ with respect to origin *e.g.*, shoulder is proximal end of the arm. Opposite to distal.

Pseudocel. Body cavity in nematodes that is not lined with membrane derived from mesoderm.

Pseudocopulation. False copulation attempted by male insect with a flower that is mistaken as female of the species due to similarity in colour, shape or scent. usually results in pollination of flower.

Pseudocyesis. Pseudopregnancy: emotionless state leading to develop!llent of pregnancy-like characters in woman even in absence of true pregnancy.

Pseudogamy. Development of ovum into new Individual due to stimulation of male gamete whose nucleus actually does not fuse with that of ovum - occurs in some species of nematodes.

Pseudopodia. Blunt temporary protoplasmic projections which are pushed out from an amoeba or amoeba-like structure during feeding or locomotion.

Pseudopupa. Insect larva in quiescent pupa-like state; one or two instars before pupal stage. common In some beetles (Coleoptera: Hexapodal.

Psittaciforms. Order of tree-dwellIng and fruit-eating birds (Carinatre: Aves) having specially articulated bill and synadactylus feet. adapted for climbing twigs and branches of trees – parakeets, parrots etc.

Psocoptera. Order of hemipteroid insects (Hexapoda), soft-bodied, biting and chewing mouthparts, long filiform antennae; wings when present have simple ventation; metariorphosis gradual – booklice etc.

Psycologist. Person having sound knowledge of psycology.

Psychology. Scientific study of behaviour especially of human beings.

Ptarmigan. Mountain-dwelling grouse. whose brown plumage become white in winter.

Pterin. Substances occurring as pigments In butterfly wings.

Pterodactyl. Extinct reptiles which were fish-eating type; flying over water and diving after (for capturing) small fishes: Body bat-like and they hung suspended from trees over hanging rocks during their time of rest - *Rhamphorhynchus* species.

Pulmocutaneous. Blood vessel (artery/vein) in frog supplying deoxygenated blood to lungs.

Pulp Cavity. Space inside the dentine of tooth and placoid scale filled with pulp (mixture of connective tissue. blood vessels. nerve fibres. mesenchyma cells called odontoblasts that secrete dentine layer).

Pulsate. To throb or beat (heart) rhythmically. *Pulsatioin*: rhythmic beating of heart.

Pulse. Blood forced into aorta during systole sets pressure wave in arterial walls as it travels. The pressure wave expands the arterial walls as it travels and this expansion is called pulse. In normal man it is 72/minute.

Pulvilus. Fleshy pad or lobe-like structure between claw pair on the terminal segment of tarsus in some insects.

Puma: Cougar or American mountain lion : large animal, *Felis concolor* (Carnivora: Mammalia) greysih-brown in colour without any spots; feeds on deer and other domestic animals.

Pupa. Development stage of holometabolous insects usually encapsulated or in coccoon.

Puparium. Hard barrel - shaped case enclosing pupa of certain insects.

Purine. Heterocyclic compound (CsH4N4) derivatives of which are found in nucleotldes. These compounds include adenine and guanine.

Purninergic. Nelve endings that release purines or their derivatives as transmitter substance.

Purkinjie Cells. One of the layers incerebellar cortex which is single layered and these cells are among biggest neurons in the body. They have very extensive dendritic arbores that extend throughout the molecular layer of cerebellar cortex. *System*: specified and specialised conducting tissue that transmits impulse from SA-node to ventricle. It exists through intelVentricular septum and termin'ates in the wall of ventricle.

Puss-moth. Large. downy-looking. white moth *Cerura vinula* (Lepidoptera: Hexapoda) of Europe whose grostesque caterpillar with its forked posterior appendage (tail) can adopt waming attitude and eject formic acid in self defence.

Pustule. Pimple: small circumscribed elevation on skin. containing pus.

Putrefaction. Destruction of organic matter by bacteria.

Pycnodontidace. Family of extinct fishes of Jurassic and Cretaceous periods that had high oval laterally compressed body covered with rhomboidal ganoid scales.

Pycnogonida. Sea-spider; long-legged marine arthropod *Pycnogonum littorale* (Arachnida) resembling spider; on coral polyps and sea-anemones.

Pycnontidce. Family of singing-birds (Passertformes: Aves) - bulbul *Pygal*: belonging to rump or postenor of animal.

Pygidium. Posterior part of abdomen in certain arthropods formed by fusion of several body segments.

Pygochord. Thick ridge-like out growth from hind gut in some hemichordata.

Pygostyle. Plough-shaped structure in birds fonned by fusion of last 3-4 caudal vertebrae around which tail feathres are arranged in a fan-like manner.

Plyangium. Proximal most muscular part of conus arteriosus in frog.

Pyloric. Belonging to plyorus or posterior end of stomach.

Pyramidal Neurons. Multipolar neutrons *i.e.,* with several dendrites and one axon. *Tract:* bundle of neive fibres originating in the motor cortex and descending down the brain stem to the modulla and spinal cord. Responsible for mediating control of voluntary muscle movements.

Pyramid of Kidney. Medulla of mammalian kidney is divided into pyramid-like areas with the help of renal columns of Bertini.

Pyrogens. Fever causing substances that act on thennoregulatory centres of hypothealamus to reset them to produce fever. They are usually proteins (endogenous pyrogens) with molecule. Weight of 10000-20000 and released from polymorph nuclear leucycytes in response to viruses. antigens, bacteria. antigen- antibodycomplexes and injuries etc.

Q

QRS-Duration. Ventricular depolarisation in ECG having a period of 0.08 econd.

Quadrate. Bone of suspensorium that helps to suspend lower jaw in authosytylic jaw suspension. (It is found in amphibia. reptiles and birds).

Bone. Cartilage bone at hind end of upper jaw in most vertebrate animals.

Quadratojugal. Small bone connecting quadrate and jugal bones on either side of skull in some vertebrate animals.

Quadripara. Woman who has undergone four deliveries.

Quadroon. Offspring of white and mulatto (quarter-negro blood)

Quadrumana. Primates other than man characterised by having both thumb and large toe opposable.

Quadruped. Four legged animals like cow, sheep, goat, elephant etc.

Quagga. Extinct South Africa. striped wild ass. *Equus quagga* (Perissodactyla: Mammalia).

Quail. Comrilonname for various small. terrestrial. grain-eating. migratory. game-birds. *Cortumix* spp. (Galliformes: Aves).

Quantal Release of nerve impulse Transmitter. Small quanta or packets of acetylcholine released randomly from nerve cell membrane at rest. Each produces a miniature end potential which is about 0.5 mV in amplitude.

Quarternary Malaria. Malaria in which the recurrence of fever occurs after every 72 hours (every fourth day). This is due to infection of *Plasmcx1ium malariae.*

Queen. Reproductive female of social insects-honey-bees, termites and ants.

Quelea. Small passerine bird. having stout red beak, lives in flocks on grasslands of Mrica. feeding on grain-crops.

Quetzal. Beautiful bird. *Pharomadws* moctno of Central American valued for its plumage.

Quantipara. Female who has borne children (dead or alive) five times.

Quinones. Group of chemical substances involved in formation of yellow and red pigments of the insects.

Quokka. Small greyish-brown wallaby (Marsupialia: Mammalia) nocturnal and herbirvorous; lives in swamps in South-east Australia.

R

Rabbit. Small herbivorous burrowing mammal, *Oryctolagu·s cuniculus* and allied species (Lepoidre: Lagomorpha). having small ears, short legs, hind legs strong adapted for leaping; fast runner (40 to 50 kin/hour) and gregarious.

Racer. Non-poisonous. slender-bodied snake (hiphidia: Reptilia) about a metre long; found in warm and dry regions. Horse. used for races.

Rachis. Distal tapering portion of shaft of feather which becomes elongated and bears barbs. barbules etc., onit. Also refers to central axis in the testes of *Ascaris* around which cells are arranged radially.

Radials. Rest of pterygiophores other than basalia (at the base) to which fin rays are attached.

Radial Canal. Canal radiating from disc of sea-stars and related animals which runs through arm and supply tube feet. Also certain canals of Jelly-fishes and fiagellated canals in certain sponges such as *Sycon.* spp.

Radial Links. Extensions from peripheral doublets to central sheath in cilia and lagella.

Radial Symmetary. Condition in which similar parts are arranged about a common centre like wheel's spokes.

Radiata. Eumetazoans such as coelentaerates with primary and secondary radial symmetry.

Radiation Organ. In pit vipers there are pits sensitive to heat radiations and are called radiation organs.

Radical Mastectomy. Operation uf breast for cancer in which axillary lymph nodes are removed.

Ranid. Family of Dlplasiocoela (Phaneroglossa: Amphibia) frogs having protrusible tongue. distinct tympanum and separate Eustachian tubes; fadopole with single spiracle on left side -Ranaspp.

Rank (Category). Position or status of taxon in classification hierarchy - phylum. dass. order. family. genus. species etc.

Ranveier's Node. Node of Ranvier structure formed by constriction of axon of nerve cells.

Raphidate. Family of Columbiformes (Aves) that includes extinct bird dodo. *Raphys cucullatus* .

Raphidioidea. Monofamilial s~perfamily of snake-files. having mandibulate mouthparts; both larva and adult predacious on other insects - Raphidiidle.

Raptorial. Living on prey: (birds) adapted for snatching and seizing the prey.

Ras (*Reticular Activativating System*). The system in brain which produces conscious. alert state. that makes perception and localisation of individual sensation possible. System occupies midventral portion of modulla and mid brain and is made up of myriads of small neurons arranged in complex intertwining nets. Localised within this are centres which regulate respiration. blood pressure. heart rate and other vegetative functions.

Rasorial. Adapted for scratching for food (Bird).

Rat. Common name of over 700 mammalion species belonging to various families of Rodentia (Mammalia); distinguished from mouse in having teeth modified for growing - *Raths* spp. are most common.

Rate Constant. Proportionality factor by which concentration of reactants in enzymatic reaction is related to reaction rate.

Ratel. Honey badger. *ivtelltvora capensis* (Carnivora : Mammalia) found in India and Africa feeds on almost anything but is fond of honey

Ratitae. Group of large iligtless running birds - emu, ostrich. cassowaries etc.

Raven. Big-crow, *Corvus corax* (Passeriformes : Aves). about 60cm long. hoarse-voiced. completely black in colour: nests In trees or on rocky ledges; feeds on carrion and small mammals: found in Eurasia and North America.

Ray. Group of marine catilaginous fishes belonging to hypotremata (Selachii : Pisces) harming flattened body, board pectoral fins and long spine like tail, mostly bottom dwellers - devil-ray, sting-way. whip-myetc

RBCs (*Red Blood Corpuscles*). Erythrocytes: speciallsed blood cells that carry oxygen because of the presence of haemoglobin in them. The are produced in red bone marrow cells (haemocytoblasts) in case of mammals in non-mammals these are formed within liver as well as spleen Rounded, oval and biconcave in mammals and biconvex in non-mammals: generally nucleated except in mammals (excluding camel where they are nucleated). Life of these cells is 100 days in non-mammals and 120 days in mammals. The number being 54 million/mm man and 4.8 μm woman. Mammalian RBCs are 7.5 μm in diameter and 2 μm thick. They contain 29 mg of haemoglobin In mammals. There are 3X 10^{13} RBC /900 gm of haemoglobin in circulating blood of man.

Reactive Hyperemla. Response of blood vessels which occurs in many organs but is visible only in skin and is seen by increased amount of blood in a region when circulation is re-establlshed after a period of occlusion.

Recepitulation Theory. Ontogeny repeats Phylogeny doctrine that the individual in its development passes through ancestral history of species.

Receptive Field. That area of a sensory field which on stimulation influences activity of a given neuron

Receptor Molecule. Molecule situated on the outer surface of cell memebrane interacting specially with messenger momecules such as hormones or transmitters.

Receptor Potential. A membrane potential change elicited in receptor cells by flow of receptor current.

Receptors. A sensory end organ device for registering changes in internal/external environment of organism .

Recessive Trait. Trait that appears only when both members of an-alletic pair of genes are alike *Le.* homozygous condition. Compare with dominant trait.

Red Eyes. Bloodshot eyes or eyes with eyclids SQre from excessive weeping.

Red Fox. Reddish-brown fox of Europe having thick fur, 75 cm long body with bushy tail another 40 cm; scavenger. feeds on ground-nesting birds, insects and vegetable matter.

Red Gland. In cods and perches the epithelium of red body becomes glandular and folded to form composite structure called red gland.

Red Marrow. See bone marrow: Active cellular marrow. Opposite to inacticve slow marrow which is called yellow marrow.

Red Meat. Beef, mutton etc. meat of goat, sheep, cow, horse etc.

Red Mullet. Marine, valuable food-fish, grows upto one metre in length: found in. tropical and temperate seas especially Mediterranean.

Red Nucleus. Structure associated with basal ganglion.

Red Panda. Cat-bear: nocturnal, carnivore mammal feeds on fruits and leaves; lives in Himalayas; valued for its rich chestnut-brown fur.

Red Poll . Small sparrow-like finch (fringillidre: Passeriformes) with crimson forehead and black chin; found in North Eurasia and North America; feeds on seeds and insects.

Red River Hog. Stout-bodied, reddish-brown pig. *Sus* spices (Artiodactyla: Mammalia) with large head and tufted ears, weighs about 1.5 quintals; found in Mrica, feeds on fruits and garbage.

Red Shank. Medium- sized wading bird, *Tringa* spices (Aves); with conspicuous red legs.

Regeneration. Replacement of a body part, by growth, which has been lost. tail of lizard, tongue of man, body of sponges, hydra etc.

Regurgitiation. Taking back something from alimentaly canal into mouth *e.g.*, birds regurgitate milk from crop into mouth and then feed it to young ones through their beak.

Reindeer. Large heavy deer, *Rangifer* spp. (Artiodactyla: Mammalia) with large palmated antlers in both sexes; used for drawing sledges and herds maintained for milk, flesh and hide.

Relaxation. Coming back of contracted or extended state to normal e.g., muscle relaxation.

Relaxin. Ovary secretion during pregnancy. A polypeptide that loosens ligament of pubic symphysis (often softens cervix) and facilitates the delivery.

Remiges Remex. Contour feathers with equal vane and found on wings or patagium. In pigeon there are 11 remiges attached to hand (primaries). 12 attached to foreerm (secondaries). Upper arm does not bear these, if at all present, they are tertiaries.

Remora. Marine fish that attaches itself to sharks, turtles and even ships by means of a sticking disc on top of its head and suck fishes.

Rem-Sleep. Rapid eye movement sleep: sleep state where eves move rapidly.

Renal. Any structure like artery/vein pertaining to kidneys. ***Blood Flow:*** Blood to renal tubul is supplied by afferent renal arteriole from renal artery which branches into Bowman's capsule to form capillary net-work called glomerulus from which arises efferent renal arteriole. The efferent renal arteriole divides into many capillaries which supply the tubule before draining into renal vein. the branches of efferent renal arteriole joins with venule that opens into renal vein.***Blood Coltimns of Bertini***: Connective tissue partitions separating medulla of kidney of mammals into a number of pyramidal areas. ***Blood Erythropoitic Factor***; Substance secreted by kidneys which form erythropoitin (which regulates RBCs level in the blood) by action with plasma globulin. Its production is increased by hypoxia. cobalt salts and androgens. ***Blood Homestasis***: Maintenance of tonicity. volume and specific ionic composition particularly H+ ions concentration of ECF by the kidneys in order to maintain constancy of internal environment. ***Blood Hypertension***: Constriction of one renal artery causing development of sustained hypertension. This is often due to increased renin secretion. ***Blood Portal System***: Portal system involving veins from posterior limbs and posterior part of the body opening into the kidney. These veins distribute the blood into kidney issue from where it is recollected from kidney that joins to form renal vein/veins that open in to pos terior vena cava. ***Blood Threshold***: Minimum concentration of substance necessary to produce a detectable response in kidney *e.g.*, renal threshold for glucose is the plasmo level at which glucose first appears in the urine in more than normal minute amounts and is about 200 mg/dl of atrial plasma which corresponds to a venous level of about 180 mg/dl. ***Blood Uremia***: Disorder of renal function where breakdown products of protein matabolism accumulate. It is characterised by anorexia. nausea and vomiting. mental deterioration and confusion. muscle twitching. convulsions and coma. Because erythropoisis is depressed anaemia is prominant feature of chronic uremia.

Renin. Proteolytic enzyme secreted by kidneys into blood stream (molecular weight approximately 40000). It acts on a glycoprotein In a 2-globulin fraction of proteins In the circulating plasma and releases a decapeptide angiotensin-I which is converted to angiotensin-II by dipeptidyl-carboxypeptidase and renin-angiotensin system regulates aldosterone secretion and regulates renal blood pressure etc. With reference to digestive enzymes it is believed to be milk coagulation enzyme responsible for milk curdling.

Respiratory Dead Space. Space occupied by gas that does not exchange with blood in pulmonary vessels.

Respiratory Organ. Breathing organ/respiratory surface : organ/ some part of body set aside for exchange of gases with blood surrounding.

Respiratory Pigment. Organic pigment that transports oxygen in the body *e.g.*, haemoglobin, haemocyanin etc. In non-chordates it is dissolved in plasma but in chordates it is present in RBCs.

Respiratory Surface. Part of body where exchange of gases (oxygen and carbon dioxide) is facilitated during breathing. It is in form of lung in case of most of terrestrial vertebrates and gill in aquatic vertebrates.

Response. Reaction to stimulus (either external or internal).

Restiform Body. Inferior peduncle connecting cerebellum with brain stem.

Resting Potential. Normal unstimulated cell membrane potential of a cell at rest.

Rete Mirabile. Extensive counter current arrangement of arterial and venous capillaries.

Rete Testes. System of tubes in testes of mammals to which proximal ends of vasa efferentia are connected.

Reticular Tissue. Fibres in this tissue are compounds of a protein called reticulin. They are branched but do not form fascia and occur as sheets forming basement membranes below epithelial tissue; found in sarcolemma around muscle fibres; neurilemma around nerve fibres and form framework of many lyphoid organs.

Reticulo Endothelial System. Tissue macrophage system including Kupffer's cells of liver and alveolar macrophages in lung coming from monocytes; concerned with protection of body against infections etc.

Retinaculum. Appendage on abdominal somite of springtails (Collembola: Hexapoda) that articulates with furcula. Clasp on morewing of certain moths (Lepidoptera) for retaining frenulum of hindwings.

Retinula. Group of pigmented elongated cells in arthropods present in basal portion of compound eye.

Receptor Muscle. Muscle that pulls back some structure opposite to protractor that pulls forward.

Retortamonadida. Order of zoomastigophora (Protozoa) that includes unicellular. flagellate parasitic organisms.

Retrpgrade Amnesia. Loss of memory for events immediately pro ceding brain concussion or electric shock therapy.

Retrogessive Metamorphosis. In certain animals like Herdmania spp ., larva has more advanced chordates' characters like notochord. tubular nerve cord and pharyngeal slits etc. which are lost during metamorphosis and adult shows an organisation totally different and that of primitive type. Such a metamorphosis is called retrogressive nletamorphosis.

Reptabe Mechanism. Some of nor-epinephrine in nerve endings is manufactured there but some is that which has been secreted and then taken up again in adrenergic neurons. This is characteristic of adrenergic neurons and is called reuptake mechanism.

Reversal Potential. Membrane potential at which current through activated synapse or receptor membrane becomes null.

Rh·Factor. Kind of antigen found in some RBCs. first found in RBCs of rhesus monkeys, hence named Rhesus factor (Rh-factor). About 97% of Indian population is Rh-positive and 3% Rh-negative. The antigen has significant role in blood transfusion and marriages of individuals. Rh+ is a dominant trait and marriage of such individual with a Rh- female would lead to sensitisation of the body and would cause agglutination of foetal cells in the pregnancies after first or maximum second one. These complications result in abortions, miscarriages etc.

Rheorecetor. Receptor responding to water current as lateral line organ in fishes.

Rhinal Process. Small cartilage closing nostrils of frog during breathing.

Rhinarium. Small moist hairless area of skin surrounding the nostrils as found in many mammals like rabbit etc.

Rhinencephalon. Anterior ventral part of the anterior end of fore brain that differentiates into olfactory lobes.

Rhinocel. Cavity often present in each olfactory lobe of brain of a vertebrate.

Rhinoceros. Large. thick-skinned. plant-eating mammal (Perissodactyla) having one (sometimes two)upright horn on snout; found in South Asia and Africa.

Rhinoceros Beetle. Dynastid beetle (Coleoptera: Hexapoda). smooth shiny reddish-black to black in colour having horn-like projection dorsally upwards from head - *Oryctes rhinoceros.*

Rhinocerotoidea. Superfamily of odd-toed hoofed mammals (Perissodactyla: Ceratomorpha) that includes all the living and extinct rhinoceroses.

Rhinolophidal. Family of small horse-shoe bats (Chiroptera: Mammalia) insectivorous. aerial animals characterised by having large ears and extremely complex hor::1e-shoe shaped nose leaves.

Rhodopsin. Retinal pigment in rod sensitive to low levels of illumination.

Ribonuclease. Pancreatic enzyme acting on RNA nucleotides. **RIBOSE** : $HOCH_2$ (CHOHb CHO: pentose monosaccharide and constituent of RNA.

Ribosomes. Ribonucleoprotein particles found in thecytoplasm: sites of protein synthesis (sites of interaction of m-RNA. t-REA and amino-acids during protein synthesis.

Ribs. Extensions of basiventralia in the sternum of animals like rabbit and other tetrapods. Two. types - ventral or pleural and true or dorsal ribs which may further be single headed or biheaded.

Rickets. Condition in young ones developed due to poor absorption of calcium as a result of calcium deficiency leading to hypocalemia and failure of protein of bone to mineralise.

Robin. Small red-breasted song-bird of Europe. *Erithacus rubecula:* red-breasted thrush of North America *Turdus* migratorius: any red-breasted song -bird of India or Australasia.

Rodent. Any species belonging to order Rodentia, gnawing mammal. *Rodentia*: order of mammals (Eutheria: Chordata); characterised by chisel-like persistently growing incisors; gait is plantgrade-rats. mice. squirrels and porcupines.

Rods. One of the nervous structures in retina having rhodopsin and sensitive to low levels of illumination.

Roe. Mass of fresh-eggs: female of small red deer, *Capreolus capreolus* (Mammalia) (Vroe-buck).

Roller. Group of highly coloured birds *Coracias* spp. (Coraciformes: Aves) characterised by rolling movement, like tumbler-pigeon, during courtship display - C. *garrulus* is most common species.

Rook. Gregarious species of crow, *Corvus frugilegus* (Passeriformes. Aves), haVing raueous voice.

Rotation. Ventral divergence of gastrula of frog about horizontal axis due to movement of cells outwards and dorsally away from midventralline.

Rotator. Type of voluntary muscle which brings about the rotatory motion of one part of organism on other.

S

Seabellidae. Family of feather-duster worms. sedentary polychaete annelids occurring in intertidal depths; comprises two subfamilies - Fabriclince and Sabellince.

Saccopharyngoidei. Suborder or actinopterygian fishes, that includes eel-like marine fishes having no swim bladder and narrow filamentors tail - gulpers.

Sacculus. Ventral part of internal ear which is connected with utriculus through a sacculo-utricular canal.

Sacculus Rotundus. Rounded sac at the junction between ileum. colon and caecum. It has definite valves which can direct food either -through caecum or into colon.

Sacculus Vasculosus. Thin walled dorsal expansion of diencephalon. It is a pressure receptor. The organ is well developed in deep sea fishes.

Sacroiliac Joint. Sacral ribs articulate with ilium bone of pelvic girdle forming this joint.

Sacrum. Region in vertebral column following lumber region and consisting of 3-4 vertebrae. This is followed by caudaJ region.

Saithe. Coalfish: sea-fish with dusky skin; which soils fingers like wet coal.

Sigittal. In a bilaterally symmetrical animal pertaining to the median antero-posterior plane.

Salamandar. Tailed amphibian (Mutabilia:Urodela); soft-bodied with moist skin, about 15 em long (including tail); larva aquatic, adult mostly terrestrial - *Ambystoma maculosa. Proteus anguinus. Siren lacertina.* etc.

Saliva. Watery, colourless secretion of salivary glands; containing water, mucous, amylase and mucopolysaccharidase. It lubricates the food, helps in digestion of starch and kills harmful bacteria.

Salivary Cells. Papillae in front of tongue and covered with stratified epithelium from which flakes are continually falling to form saliva cells.

Salivary Glands. Glands within mouth cavity which secrete saliva.

Salmon. Common name of several sea-fishes, *Salmo* spp (especially *S.salar)* with silvery sides, ascends rivers or spawn; highly esteemed food fish.

Salmonidie. Family of suborder Salmonoidei (Salmoniformes: Pisces) includes soft-rayed fishes like salmons, trouts etc.

Salt Gland. Birds like herrings, penguines and albatroses liVing at sea and *very* far from land have a pair of glands in the head which secrete salts to maintain osmoregulation.

Sandfly. Small Dipteran insect *Phlebotomus* spp.(Hexapoda); having biting mouthparts, acts as vector for transmitting phelebotomus (sandfly) fever.

Sand Lizard. Common name for *Uromastix herdwickii.*

Sapajou. Common name for New World monkeys - capuchin monkey *Cebus* spp. have black hair like a cowl at back of head and prehensile tail:spider monkey *Ateles* spp. with long slender legs and prehensile tall.

Saurpterygia. Order of fossil reptiles incluaed aquatic and amphibious forms - plesiosaurus etc.

Saw-fish. Common name for a cartilagenous fish *Pristitis* species, which has a saw-like structure formed by elongation of ethmoid region of skull.

Sawfly. Hymenopterous insect (not a true fly as it has four wings). saw-like ovipositor for cutting plant tissue to lay eggs therein -*Athalia lugens proxima.*

Scala Media. Cochlear duct itself in mammals. It contains endolymph and this is situated in basilar membrane along the floor. Over basilar membrane is found receptor apparatus as organ of Corti.

Scala Tympani. Lower tubular portion of cochlea in mammals filled with perilymph. Dorsal to this is scala media and separated from it through basilar membrane.

Scales. Skin derivatives mainly found in fishes and reptiles forming exoskelton. Reptilian scales are epidermal and dry while those of fishes being both ecodermal and endodermal and have pulp inside (within the pulp cavity).

Scallop. Marine bivalve mollusc *Pecton maxtmus* (Anisomyaria Pelecypoda) with fan-shaped shell having ridges radiating from middle of hinge and edged with small rounded lobes

Scalyant-Eater. Conlmon name for *Manis* Spp (mammals).

Scaphopoda. Tooth-shelled molluscan class of non-chordates.

Scapula. Shoulder-blade in vertebrate. Bone in the half of pectoral girdle.

Sapcoptidae. Family of Acarina (Arachnida:Arthropoda); that includes mites, causing skin diseases like itch, mange and scabies to man and other animals like dogs, hogs, horses, sheep *etc.-Psoroptes equi ovis* etc.

Scizocoel. Coelom developed from originally solid mesoderm and differentiated on the sides of notochord (also called lateral plate mesoderm) as found in chick. It differs as compared with eterocoelic coelom in which mesodermal pouches are formed on both sides of alimentary canal.

Schiziphrenia. Serious mental disorder characterised by a withdrawal from normal relationships into a personal world of victim's making.

Schwann Cell. Neuroglial cell that wraps its membrane around axons during development to produce a myelinated insulating sheath between nodes of Ranvier.

Scientific Name Method. Method involving primarily the formulation of hypothesis on the basis of a relatively small amount of knowledge, then testing correctness of hypothesis by securing more facts by observation. arranging the facts to determine the inter-relationships and often drawing logical conclusions.

Sclerites. Hardened body wall plaited bpunded by sutures as seen in arthropods.

Scleroblast. Amoebocyte like cell in sponges which secretes spicules. It is termed calcoblast if secreting calcareous spicules. siliceoblast if secreting siliceous spicules and spongioblast if secreting spongin fibres.

Sclerotic Coal (Sclera). Outside layer white of an eye.

Sclerotic Ossicle. Overlapping ossicle-like bodies in sclerotic coat of the eye of reptiles.

Sea Butterfly. Small free-swimming, gastropod mollusc.

Sea Cake. Disc-like echinoid, Echinarachnius species.

Sea Canary. White whale (Cetacea: Mammalia).

Sea Cat. Common name for various cat fishes (Siluridre :Pisces).

Sea Cow. Common name for manatees: aquatic mammal (Sirenia).

Sea Crow. Common name for various aquatic birds including chough, cormorant etc.

Sea Dog. Sea-hound: dog-fishes and small sharks.

Sea Duck. Red-headed diving-duck *Nyrocafertnc*..

Sea Elephant. Elephant seal, *Mirounga* species.

Sea Ginger. Hydrozoan coral (Coelenterata) with many polyp-cells.

Sea Gooseberry. Common name for comb-jellies with spherical bodies and belonging to phylum ctenophora.

Sea Grapes. Egg-masses of cuttle-fish. *Sepia* species.

Sea Gull. Long-winged, web-footed sea bird (Laridre).

Sea Hare. Common name for mollusc *Aplysia.*

Sea Hawk. Keen -sighte. diurnal,· aquatic, predatory brid. *Stercorarius skua* (pelecaniformes: Aves).

Sea Hog. Sea-pig: short-snouted dolfin.

Sea Horse. Small tailed fish *Hippocampus* species.

Sea Jelly. Jelly-fish (Coelenterata), marine, free-living animal.

Sea Leopard. Spotted seal (Carnivora: Mammalia).

Sea Lily. Featherstar : small maline echinoderm, *Antedon* species.

Sea lion. Large seal, aquatic mammal belonging to order ennipedia of eutherian mammals.

Sea Mew. Nickname for various sea-gulls especially *Larus* species.

Sea Monster. Giant squid. highly developed marine mollusc.

Sea Mouse. Common name for annelid *Aphrodite* spp.

Sea Owl. Lumpsucker: spiny-finned sea-fish, *Cyclopterus* spp., having pectroal fins modifed into sucking discs.

Sea Pansy. Common name for members of Pinnatulacea, an anthozoan cnidcuian.

Sea Parrot. Puffiun brightly CQloured sea-birds. *Fratercula* supp: having parrot-like beak.

Sea Sleeve. Cuttlefish. *Sepia* spp. (Decopoda.: Mollusca).

Sea Snake. Common name for group of snakes (Hydrophidre), compressed tail and venom. dangerous to human beings.

Sea Spider. Common name for arachnid *Nymphon* spp. belonging to class ychgonida

Sea Star. Common name for asteroid echinoderm star-fish.

Sea Swallow. Aquatic fowl with long pointed wings and forked tail.

Seal. Amphibious furred carnivous mammal *Phoec Vvtultna* having tapering body like fish and broad flippers; lives below ice surface.

Seaceous Glands. Glands consisting of few small sacs alveoli secreting an oily substance called sebum (formed by estesr of cholesterol) and associated with hair follicle. They are also found in non-hairy areas like glans of penis, cornea of mouth and eye. internal surface of prepuce. lips of vagina etc. Their secretion lubricates hatr and skin.

Semi Conservative Replication. Synthesis of two new DNA molecules ut1lising each of the two single DNA strands composing the original molecule as a template for the assembly of complementary strands.

Seminal. Structure pertaining to supermatozoa. May be in form of either seminal receptacle which is a sac-like organ that receives and stores sperms after their release in most animals and considered to be a part of female genital system (receptacula seminalis) or it may be in form of seminal vesicle which is also a sac-like organ in several animals in which seminal fluid is stored before being discharged and considered to ,be a fart of male genital system.

Sensory Epithelum. Epithelium specialised to receive stimuli, to conduct impulses etc. Such cells are found in receptors derived from surface *e.g.*, retina of eye.

Secsory Homunculus. Representation of localisation of fibres from various parts of body in posterior central gyrus, with legs on top and head at the foot of gyrus. Cortical areas from trunk and back are small whereas very large areas are concerned with impulses from head and parts of mouth related to speech.

Sensory Neurons. Neurons in which impulses travel from a receptor system as for example skin to brain or spinal cord. Also called afferent neurons. Cf. motor or efferent neurons.

Septomaxillary. Small bones covering each nasal capsule in many animals like frog.

Septum. Partition that separates two cavities or structures.

Septum Mediale. Median septum in frog's heart separating aperture leading to left systemic arch.

Sere. Set of communities formed during ecological succession in an 'area that ultimately lead to stabilisation of area from extreme, conditions. -Sere consisls of one pioneer, many transitional and one climax.

Serial Homology. Occurrence of homologus structures in different segments of the same individual. Homology repeated in a series.

Serosal. Pertaining to the side of an epithelial tissue facing the blood as opposed to mucosal side which faces the exterior or luminal space.

Serpentes. Common name for various reptiles which are limbless and popularly known as snakes. They may be poisonous or non-poibnous.

Sertoli Cells. In vertebrates the non gametic cells in the seminiferous tubules which seem to provide nourishment for developing sperms and also secrete inhibin.

Servo Mechanism. Control system that utllises negative feed-back to correct derivations from a selected level, the set point.

Sex. Sum of characters that distinguish male and female of a species.

Sex Cells. Gametes produced by gonads (testes/ovaries). Formation is known as gametogenesis in general and spermatogenesis or ovogenesis for sperm or ova formation respectively.

Sex Chromosomes. X and Y chromosomes specially concerned with sex determination.

Sex Determination. Genetical basis of determining sex of individual in a dioecious animals like human beings. The male individuals have XX chromosomes and female individuals have XX chrumosomes. This sex determination is XY type. The other type of sex determination may be XO found in other animals.

Sex Hormones. Hormones secreted in the body that influence sexual functions and. secondary sexual characters in the individuals. These hormones are secreted by interstitial cells of gonads *(Le.,* Leydig cells in males and copus luteum in females in case of human beings) and are called primary sex hormones. The hormones secreted from other places like placenta and adrenals etc., are called secondary sex hormones. These influence the development of embryo in the desired directions (according to genetic constititution XX/XY), guide development of child into boy or girl, bring about puberty and ensure reproductive function, guide physical/ physiological/behavioural changes in Males/females.

Sex Limited Characters. Characters belonging to only one sex. They do not find their expression in opposite sex because the expression is dependent on a particular and specific honnone *e.g.* baldness is a character that finds its expression only when testesterone honnone is present and for this reason it is found only in males.

Sex Linked Characters. Characters with their linked genes located in sex chromosome a number of such characters like haemophilia, night blindness, colour blindness etc., are linked to X-chromosomes while characters like hypertrichosis are liked to Y'chromosomes. These characters generally make their expression in males because of heterozygous genetic constitution but rarely make their appearance in females because a female requires double dose of such X-chromosomes which is a very rare incidence.

Sexless. Without sex: having no sexual feelings.

Sexual Cycle. Formation of offspring from parentai organisms by process involving sexual reproduction.

Sex Dimorphism. Differentiation of two sexes by external characteristics in case of dioecious individuals, *e.g.,* male frog has copulatoxy pads, vocal cords but female frog lacks these characters.

Sex Play. Temporary mating of male and female of same species without transfer of sperms into female reproductive tract.

Sex Reproduction. Process by which a vegetitive body is formed from another vegetative body, involving fusion of sex cells or gametes.

Seymouria. Lizard-like early reptile of Carboniferous period (now extinct) that probably developed from amphibians and started laying cleidoic (shelled) eggs on land; it was about 60-80 cm long with short study pentadactyl fore- and hind limbs, fed on insects and other small creatures.

Shaft. Slender, middle portion of a long bone as femur. Also middle' keratinised portion of avian feather and the portion of hair projecting outside the body in mammalian hair.

Shark. Long elongated fish with endoskeleton of cartilage and exoskeleton of dermal scales; mostly marine, some ascend the rivers; rough skin, blunt snout, ventrally placed mouth and powerful jaws; voracious feeders - *Scoliondon dumerilli.* S. *palasarrah,* S. *Sorrakowah* and S. *walbehmi* are found in Indian seas; *Carcharhimus* spp. In rivers.

Sheath. Membranous covering of muscle, nerve or blood vessel: prepuCe of male animal.

Sheep. Gregatious wooly horned ruminant, *Ovis* ammon and allied species (Artiodactyla: Mammalia), wild or domesticated tame, bred for flesh and wool.

Shell. Outermost protective covering made up of mainly Inorganic material *e.g.,* calcium carbonates silica, often chitinous or pseudochitinous.

Sheep Glands. Glands mainly within reproductive system which secrete shell over the egg.

Shipworm. Group of bivalve mollniscs (Teredindrel that look like earthwonps but are usually found boring wood, especially of bigboats.

Shiverings. Involuntary rhythmic trembling of body during cold, fear or excitement. The basic purpose of the mechanism is to produce large amount of metabolic energy in order to cope with stressful situation *e.g.,* during cold this energy is used in the individual to raise the body temperature to cope up with the heat loss and thus helps in homeothermy during fear this energy helps Individual to be prepared for necessities of fight or flight reaction.

Short Adrenergic Neurons. Myometrium of uterus unlike the rest of organs is inervated by a special system of neurons called short adrenergic neurons with cell bodies in uterus and postganglionic fibres to these postganglionic neurons presumably go all the way to the uterus.

Shrew. Small. mouse-like. insectivore mammal. *Sorex* and allied species having pointed snout and short fur.

Shrike. Butcher-bird. *Lanius* and allied species; characterised by strong hooked-bill. preying on insects. mice and small birds.

Shrimp. Small. 20 to 25 cm.1ong. marine, decapod crustacean. *Crangon* and allied species with comparatively long tail; valued as food *-c. vulgaris* inhabits sand -coasts.

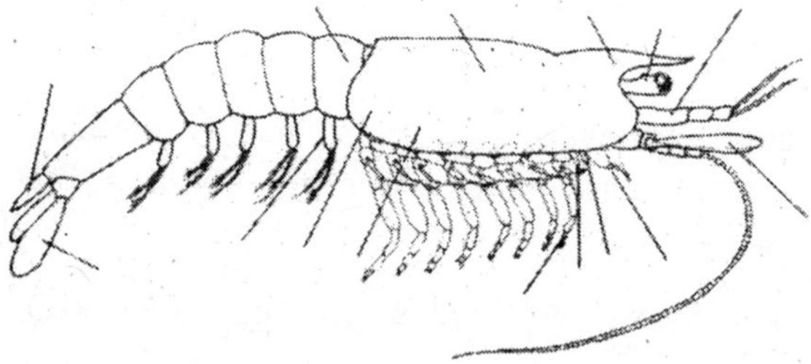

Sign Stimulus. Most basic essential pattern of sensory Input required to release an instinctual pattern of behaviour.

Silent Mutation. Point mutation which results in a change from codon-spedif)ring an amino-acid to other codonspecific for same amino-acid.

Silkworms. Caterpillars of *Bombyx* mort and allied species, (Bombycire: Lepidoptera) that feed on mulberry Leaves. spin silken cacoons that are of commercial value in silk industry.

Silver Fish. Not fish but silver-grey coloured, very active, swift running Wingless insect. *Leptsma* spp. trhysanura:. Hexapoda). feeds on starchy material like clothese, paper etc. cominon household pest.

Silvered Fox. American red fox, *Vulpes utilpes* (Carnivora: Mammalia) having silvery-tipped black fur. valued for its fur.

Simple Cells. Receptive fields of neurons in lateral geniculate body on the basis of characteristics of their responses have been classified into simple cells which respond to slits of light dearer lines or edges but a particular cell responds only when stimulus has peculiar orientation.

Sinus Glands. In some arthropods, the gland at base of eye stalk controlling chromatophore dispersion, under influence of X-gland of brain.

Sinus Venosus. Large chamber receiving blood from veins and passing it to auricle (usually right auricle where the auricles are divided) through sinuatrial opening which is guarded by a valve; absent in avian and mammalian heart (present in fishes. amphibia, reptiles etc.

Siphonaptera. Order of secondary wingless insects (Hexapoda) having laterally compressed body with hard smooth exloskeleton, antennae short, eyes absent, ocell1 present. mouthparts sucking type, metamorphosis complete, mostly ecoparasites of warm-blooded animals including man - fleas.

Siphonet. Honey-tube of aphids (Homoptera: Hexapoda).

Siphuncle. Gas-filled canal that passes through coiled and chambered shell of *Nautilus* species (cephalopod mollusc).

Siren. Mythological fabulous creature with woman's head and bird's body; living on rocky isle; lured seafarers by their singing, Eel-like. tailed amphibian (Sirenidce) short fore-limbs and no hind limbs; retains external gill throughout the life.

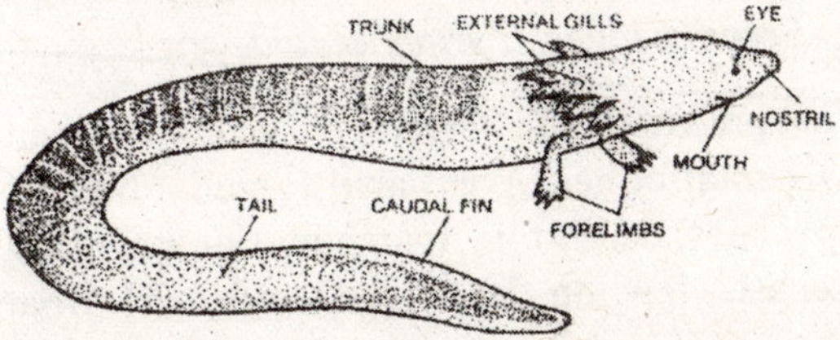

Sirenia. Order of mammals (Eutheria: Chorda) ihat includes large aquatic herbivores, with thick blubber and no hair, forelimbs modified into paddles tail with horizontal fin; males have tusks -sea-cows (manatee) etc.

Skate. Maline ray-fish. *Raja batis* (RaJidre) about one metre long with broad and rhombic disc formed by pectoral fins extended on either side. caudal fin rudimentary. males have copulatory claspers.

Skeleton. External/internal hardened frame work of animal body giving support and protection to soft parts of body. The exoskeleton includes usually epidermal derivatives like scales. feathers; hair. Nails, claws. Horns, etc.. while endoskeleton consists of an axial skeleton that includes skull and veretebral column and appendicular skeleton that includes skeleton of girdles and limbs. Endoskeleton is dermal and may be cartilagenous or bony or made up of both.

Skimkmers. Long-winged marine birds. *Rynchops* spp. (Rhychopidre) having knife-like lower mandible that is longer than upper wider mandible; found in tropical waters around the world.

Skin. Outermost investment over the body of an organism; made up of epidermis which is derived from ectoderm and dermis derived from mesoderm. Epidermis may be satratified but dermis is always composite layer. Mayor may not have glands and other derivatives.

Skippers. Small butterflies (Hespeliidre: Lepidoptera); broad head and body; mostly sombrely coloured; make short darting flights; caterpillars pests of many economic crops - *Pamara* spp; *Erynnis tages*. *Pyrgus malvae* etc.

Skua. Large rapacious predatory sea-bird *Stercorarius skua* and allied species (Charadiformes: Aves) akin to gulls; hooked beak and sharp claws; largest is found in Europe along North Atlantic coasts; feed on fish which they snatch from other sea birds.

Skunk. Small North American musteline Mephitis. *Mephitis* species. about 75 cm long. black striped with white and a bushy tail (another 40 - 50 cm) emits powerful offensive odour from two anal glands located under its tail; valued for its fur; adapted as a pet after removing the scent glands.

Slime Glands. Mucous producing glands in skin of individuals which produce slimy substance *e.g.*, mucous glands in skin of fishes and amphibia. The secretion of these glands protects skin from effect of water, keeps it free of bacterial action and males body slippery to escape from grip of an enemy.

Sloth Bear. Common shaggy blackhairded bear, *Melw"Sus* ursinus of India, feeds on insects, fruits and honey.

Smegma. Seoaceous secretion that accumulates around glans of penis and clitoris.

Smooth Muscles. Muscles without sarcomeres and hence do not carry striations. Myofilaments are not uniformly distributed within small uninucleated spindle-shaped cell. Also called involuntary or unstiped muscle *e.g.*, muscles in urinary bladder.

Snail. Common name for gasteropod mollusc. fresh-water. marine and land-dwelling forms; all have well-developed coiled shell and muscular foot for locomotion, some species are edible.

Snake. Serpent: reptile without fore and hind limb- cobra

Sneezing. Response to irritation of receptors in mucosa of large respiratory passages. It is an expiratory effort with continuously open glottis to expel irritants in nasal chamber and keep air-ways open and clear.

Snipe. Common name for short-legged wading birds (Charadrifonnes: Aves) having long straight flexible beak and mottled brown plumage; found in wet places or marshy lands. use beak to probe for in sects and worms in the mud.

Snow. Leopard ounce; *Panthera* uncia, rare carnivorous inanimmal of Central Asia. particularly Himalayas.

Snowy Owl. Large white owl (Strigiformes: Aves) of treeless tundra; males completely white and females with dark spots; feathers extend down the legs upto claws.

Sodium Pump. Membrane mechanism responsible for active extrusion of Na+ ions from cells at the expense of metabolic energy. In some sodium pum there is 3 : 2 exchange of intaceuular sodium for extracellular K+ ions.

Sole-Horn. Sub unguis below nail-plate at the tip of the digit; it is completely covered with nail-plate in a hoof. but is only partially covered in case of claws and nails.

Solitary. Living alone: not member of a colony or a group. SOLPUGIDA: order of Arachnida; small spider-like predatoxy. non- poisonous creatures with large chelicera for seizing and crushing its prey - sun-spiders.

Somatic Sensory. Afferent fibres running between cutaneous sense organs and brain/spinal cord. They receive external stimuli.

Somatodin. Factor/growth hormone acting on cartilage's linear growth and other effects. Three polypeptides with somatodin activity have been recognised (somatodin-A. somatodin-B and somatodin-C). Somatodin has insulin like properties including inhibition of lipolysis.

Somatopleur. Outer wall of splanchni mesoderm together with ectoderm forms somatopleur.

Somatostatin. Hypothalamic inhibiting hormone exercising its action on anterior pituitary secretion and inhibits extra growth. Also called growth inhibiting factor or growth inhibiting hormone.

Somites. One of the serial segments or metameres of metameric segmented animal. Also an organ or organ rudiment.

Spasm. Convulsive. involuntary muscular contraction.

Spasticity. Condition of rigidity or spasm.

Spawn. Egg -mass forming jelly-like irregularly shapes floating as in case of frog fish etc.

Speciation. Evolutionary changes in a species (characteristic usually highly correlated with environmental pressures). When complete, this process results in the fonnation of a new species.

Sperm. Male sex-cell produced by testes. Generally has an anterior head containing male pronucleus and a long postelior tail.

Spermateleosis. Process of metamorphosis of spermatids into spermatozoa: spermiogenesis.

Spermatin. Albuminoid material occuring in semen.

Spermatocytes. Cells formed by division of spermatogonia primary spermatocytes formed by mitotic division, and secondary spermatocytes by meiotic division of primary spermatocytes.

Spermatophore. Packet of sperms transferred by male to female as In insects annelids and some vertebrates.

Spermatozoa. Mature sperms formed as, a result of spermatogenesis.

Spermophile. Ground sqUirrel *Spermophilus* spp. (Rodentia: Mammalia); rodent akin to true-squirrel.

Sperm-Sacs. Pair of blind club-shaped sac like structure attached with ventral wall of urinogenital sac in male *Scoltcidon* spp and contain sea water. The water discharges from these sacs help to flush out the sperms from cloaca of male.

Sperm Whale. Aggressive toothed whale, *Physeter catadon* (Odontoceti : Cetacea), found in warm oceans produces embergris and contains mixture of spermaceti and oil in large nasal cacity.

Spenethmoid. Bone of cranium formed by fusion of sphenoid bone with etbmoid; replaces bone developed around front part of brain.

Sphenisciformes. Order of aquatic birds (Impennae: Aves) distinguished by erect posture, wings modifled into paddles and feathers scale-like; found only In southern hemisphere -pengutns.

Sphincter. Ring of smooth muscle surrounding a tube or opening by which Its contraction closes the lumen.

Spicule. One of the many solid structures that compose the skeleton of a sponge.

Spicule Penial. In male Ascaris, a chitinous rod (of which there are two) extending from cloacal opening. Used In copulation.

Spider. Small ovoid archnid (Araneida) having four pairs of legs: spin webs for capturtng its prey (insects).

Spider Catcher. Nickname of various birds (Aves) that catch and eat spiders - East Indian sun-birds

Spider Crab. Common name of long slender-legged crabs *Libtnta* spp. (Oxyrhyncha: Crustacea).

Spinal Canal. Central hollow passage running throughout vertebral column and col). tainlng spinal cord.

Crab Column. Series of vertebrae in the vertebrates which enclose the spinal cord. Also called vertebral column.

Crab Cord. Part of nervous system extending along the length of vertebral column in the vertebrates which continues anteriorly Into brain.

Crab Nerves. Nerves given out from spinal cord In the different body segments forming reflex arc from which arise ramus dorsalis. ramus ventralis and ramus communicans.

Spindle. Structure fonned within the cytoplasm during mitosis appearing to be made of fibres (asters) arranged in fonn of an elliptical structure. With this chromosomes are attached in its equatorial part.

Spine. Back-bone of vertebrate animals: quill of porcupine: immovable, sharp pointed, multicellular external process of body wal!: pointed spike in fishes.

Spinnerets. Small structures present at the abdomen of spiders infonn of finger-like organs having tiny tubes at their bases. The fluid-like secretion from these structures harden in contact with air to fonn silk threads (fonning spider web).

Spinule. Small projections in the cuticle of parasitic flat-wonns like *Fasciola* which help in anchoring the body with in the body of host and do not let the parasite being washp.d away.

Spiny-Ant-Eater. Common name for 'egg laying mammal *Tachyglossus aculeatus* (Echidna), confined to Australia having spines on skin and mouth drawn into a pointed snout suitable for ticking and eating ants.

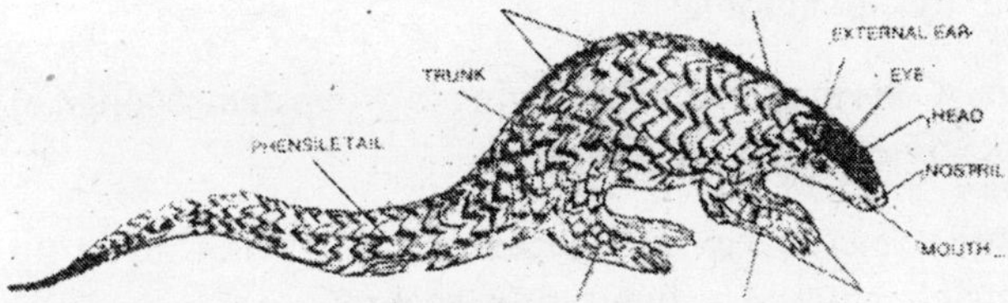

Spirochete. Long ciliated microscopic spiral-shaped organisms which may be pathogenic *e.g.,* syphlis producing siprochete *Treponema pallidum.*

Splanchnocoel. Space in the hypomere (ventral portion of enteric mesodermal pouch).

Spanchnocranium. Mesenchyme surrounding gut gives rise to series of paired arches which are called visceral arches. These arches collectively form splanchnocranium. The number of these arches is seven paIrs in most vertebrates of which the first pair forms mandibular arch, the second pair fonns hyoidean arch and 3-7 pair of arches form branchial arches. Role of splanchnocranium is to suspend the jaw with the cranium.

Spleen. Large blood-forming lymphoid organ characteristic of most vertebrates lying near the stomach.

Splenial Bone. Small flattened bone on the inner face of dentary of frog. reptiles and birds.

Splint Bone. Reduced 2nd and 5th metatarsals in horse associated with proximal part of canon bone (3rd and 4th metatarsals united to form a long bone).

Sponges. Group of diploblastic animals (Pori-fera) with porous body. central cavity (spongocoel) and chanocyte cells and regular current of water through the body. They may/may not have skeleton of spicules. They have high power of regeneration.

Spongin. Chemical present in fibres of porlferan skeleton; secreted by spongioblasts.

Spongioblasts. Cells in sponges (belonging to group demospongiae) which secrete spong in fibres.

Spongy Tissue. Tissue having spaces and not being made up of compact material - lungs of vertebrates.

Spoonbill. Large bird *Platalealeucordia* (Pelicaniforms: Aves) inhabiting lakes and marshes; large. spoon or spatula-shaped. black and yellow beak; snow white feathers and long neck; walks in shallow water preying upon frogs. molluscs. insects as also vegetation growing around.

Spores. Special reproductive bodies of lower organisms. produced generally in unfavourable conditions and provided with protective coverings and capable of developing into a new individual.

Sporocyst. Larval stage in life spp. cycle of *Fasciola* and other flukes developing from miracidium larva and giving rise to numerous rediae.

Sporogony. Sporulation: multiple fission in zygote leading to the formation of spores.

Sporulation. Process of spore formation: reproduction by asexual multiplication.

Springbok. Long-legged antelope of South Africa. *Antidorcas marsupialis* (Artiodactyla: Mammalia). able to leap as high as 3 to 4 metres in air. when disturbed.

Spiringtail. Small Collembolan insect (Hexapoda) having muscular springing organ beneath its abdomen which helps the insect to leap into air.

Stag Beetle. Large beetle, *Lucanus* cerous and allied species of Lucanide (Scarabaeoidea: Coleoptera) characterised by male having antler-like mandibles.

Stagnanthypoxia. Oxygen deficiency to snow circulation - a problem in organs such as kidneys and heart during shock. Liver and possibly brain are damaged by stagnant hypoxia in congestive heart failure.

Stallion. Uncastrated horse, *Equus* species (Perissodactyla: Mammalia) especially reared for breeding purpose.

Stapedial Plate. Columella communicates with fenestra ovalis by means of cartilagenous nodule called .stapedial plate (a detached part of otic capsule).

Stapes. Third ear bone attached to incus and oval window of cochlea.

Starch. Polysaccharide of plant origin having formula as (C 6 H 1 00 5) n . Digested by amylase to form dtand monosaccharides which are used as food in the body. It is homocyc1lc polysaccha-ride.

Star-fish. Common name for echinoderm *Pentacesors* or *Asterias* species.

Steatorrhea. Syndrome of malabsorption from intestine and characterised by passage of pale, bulky and greasy stools.

Stegocephalia. Group of fossil tailed amphibia with characters of both fishes and amphibians.

Stick Insect. Thin long apterous insect (Bacillid::e: Phasmida); stick-like, eyclindricalbody, long antennre and long legs, emits foul-smelling fluid from thoracic glands as means of defence.

Stigma. Sensitive pigme!lted eye spot in certain protozoa like *Euglena.* species.

Stoat. Carnivore mammal, slender lithe body and short legs; during summer its fur is chestnut-brown above and white beneath, but in winter it turns completely white except for black tip of tail and then it is known as ermine.

Stolon. Branch upon which colonial individual may bud upright. STOMACH: bac-like/sac-like division of alimentary canal In which food is stored and digested.

Stomatopoda. Only order of Eumalacostraca (Hopolcardia: Crustacea) having raptorial arms especially second pair of maxillipeds - *Gonodactylus* species, *Squilla* species.

Stomochord. Notochord like structure which also refers to buccal diverticulum as found in hemichordates like *Balanoglossus.*

Stomodedum. Portion of mouth cavity lined with ectoderm.

Stone Canal. Canal or passage that has calcified wall as found in various echinoderms.

Stone Fish. Ugliest and poisonous fish of Indian and Pacific oceans; body scaleless covered in warts and 13 poisonous spines on dorsum.

Stork. Large, long-legged wading bird, *Ciconia alba* and allied species (Ciconiidre Aves) characterised by long red beak;

make noise by clapping the beaks; feeds on fishes, frogs, lizards ana snakes.

Striae. Streaks. stripes. narrow bands. lines which appear specially on abdomen as a result of stretching of skin in pregnancy due rupture of lower layer of dermis. These are red at first then become silvery white.

Stratum Corneum. Outennost layer of flattened keratinised cells of epidennis skin in amphibians and mammals.

Stratum Granulosum. Cells .in epidermis fonning a layer of perfectly granular cells lying outside and above stratum spmosum. This is Just below stratum lucidum.

Stratum Laxum. Outer layer of dennis in cartilagenous fishes nearly free of fibres.

Stratum Spongiosum. Layer of cells of epidermis in malllmallying Just above stratum malpighi. Cells of this layer have small keratohyaline granules.

Strepsiptera. Order of small insects (Hexapoda} having degenerated biting mouthparts. conspicuous antennae large metathorax, metamorphosis complete; mostly endoparasites of bees, leaf-hoppers and wasps - stylopids.

Stria Vascularis. Vascular tissue over external wall of scala media. It secretes endolymph.

Strigiformes. Order of nocturnal carnivorous birds (Carinatae: Aves}, having specially culved beak and claws, well developed sound perception and noiseless flight - all owls.

Straobilation. Budding in segments of sessile scyphozoan coelenterate and larvae resulting in fonnation of free swimming plate like ephyrallarvae.

Stroke Volume. Amount of blood pumped out of each ventricle per beat is about 80 ml in a resting man of average size in supine position (80 ml from left and 80 ml from right with two ventricular pumps in series).

Stroma. Central connective tissue of ovary in mammals. In this portion the cells budded off from germinal epithelium are retained and surrounded by a group of cells that fonn Graafian follicle (it is in this structure the ovum in matured).

Struthionioal. Single family of avian order Struthionifonnes comprising single living species of ostrich *(Stnlthto camelus)*.

Struthioniformes. Monofamilial order of ratite birds (Palaeognathae: Neornithes) that includes ostriches.

Sturgeon. Common name for bony-fish *Asctpenser* - spp

Stylopodium. Division of limbs that forms upper arm in fore limb and thigh in hind limb with skeletal support of humerus and femur respectively.

Subclavian. Upper the collar bone (clavicle); refers to a particular muscle, blood vessel or other structure.

Subclavius. Small muscle producing upstroke of wing. Also called supracoracoideus or pectoralis minor.

Subcommissural Organ. Among five areas collectively called circumventticular organ the area outside the brain when an acidic dye is injected into an animal .

Subcutis. Loose, vascular and fatty layer of loose connective tissue and smooth muscle fibres in fishes like *Scoltodon* species.

Sublethal Gene. Lethal gene which affects the carrier between birth and attainment of reproductive age.

Submandibular. Salivary glands in mammals opening in the posterior part of lower jaw near the angles of jaws (its duct in man is called Wharton's duct).

Submucosa. It is usually thickest layer composed of dense elastic connective tissue-fibres with scattered fatty tissue. Fine capillaries of blood, lymph and nerve fibres traverse the submucosa.

Subspecies. Name given to geographically defined aggregate of local population which differ taxonomically from others. but is capable of interbreeding.

Substantia Gelatinosa. Lightly staining area that caps each dorsal gray columns or lemniscal system of brain to which collaterals from fibres, that enter dorsal columns, pass.

Substrate. Substance acted upon by a ferment or an enzyme.

Subthalamic Nucleus. Functionally related element of basal ganglia. Also called body of Luys.

Subumbrella. Concave, oral surface of medusa of coelenterate like *Aurelia* and *Obelia* spp.

Surface Tension. Elasticity of the surface of a substance (particularly fluid) which tends to reduce Ule surface area at each interface.

Surfactant. Surface active substance that tends to reduce surface tension.

Suspensory Ligament. Zonule: elastic suspensory structure lens; composed of radially arranged fibres.

Suture. Special fibrous joint occurring exclusively in skull. It is the tightest type of joint.

Swallowing. Deglutition: passing of food (ingested one) from mouth to pharynx and then to oesophagus.

Sweat Gkabds. Characteristic long tubular glands which have ecrine or apocrine nature and lying deep in the dermis of mammals Inner coUed portion of glands is secretory and secretes a substance called sweat. This substance is passed out of body through a duct. The glands help in regulating the temperature of body.

Swim Bladder. Air bladder: characteristic of teleosts lying just below gut outside the coelom and dorsal to gut, with shiny white balls. Consists of two chambers which are connected by a narrow duct called ductus communnicans controlled by a sphincter. May be connected with oesophagus by a pneumatic duct in physostomous fishes and having no such connection in physoclistous fishes. It has hydrostatic function.

Swimmeret. Abdominal paired appendages as found In cray-fish that function as swimmtng organs, and for carrying eggs.

Swallowtail. Large, conspicuously coloured butterflies (Lepidoptera: Hexapoda), wing span about 100 mm, with margin of hind wtngs prolonged into taillike process - *papato* spp.

Swimming Birds. Large number of aquatic birds belonging to Charadriiformes and Pelacaniformes orders of Aves.

Swine. Common name for various non-ruminent hoofed mammals belonging to family Suidre.

Swinery. Piggery: place where swtnes are kept for rearing.

Sycon. Canal system in sponges (Porifera).

Syllid. Family of free-swimming polychrete worms (Annelida) having long, linear, translucent, body and uniramous para podia.

Sylopidae. Family of beetles (Coleoptera: Hexapoda) elytra of males reduced to small leathery flaps while hind wings are large and fan-shaped.

Symbiosis. Living together/existence between two different species in which both are useful to each other *e.g.*, presence of *Zoochlorallae* alga in the body of *Hydra* where the alga gets protection from the later and the oxygen produced by alga is used by the *Hydra.* Cf. parasitism.

Sympathetic Nervous System. Portion of autonomic nervous system composecl of a chain of ganglia on the both sides of vertebral column in thoracolumber region and sending fibres to all smooth muscle tissues like heart. lung. alimentary canal etc.

Symphyla. Subclass of Myrtapoda that includes tiny centipede-like creatures found inhabiting humus or soil - garden centipede.

Symphysis. Union of bones by cartilage or fusion: immovable articulation of bones connected by fibro cartilaginous pads.

Symplectic. In bony fishes a small bone derived from hWlmandibular and attached to quadrate in methystylic jaw suspension.

Synangium. Peripheral muscular portion of conus arteriosus in lower vertebrates animal like frog.

Synapse. Conjugation between two directly interacting nerve cells where impulses in the presynaptic cell influence the activity of postsynaptic cell.

Synapsida. Subclass of Reptl1la (long extinct) that had both reptilian and mammalian characteristics.

Synapsis. In meiotic cell division the pairing of homologous chromosomes during zygotene stage (they come in contact with each other so that the number of chromosomes in the dividing cell looks apparently reduced to half).

Synaptic Cleft. Space normally more than 200 Aacross which transmitting chemicals move during neural transmission.

Synaptic Delay. Charatertstic time lag encountered from the time on impluses reaches a presynaptic nerve terminal to the time a postsynaptic potential change occurs.

Synaptic Efficancy. Effectiveness of a presynaptic impluse in producing a posynaptic potential change.

Synaptic Knobs. Terminal buttons: structures at the tip of axons containing synaptic vesicles.

Synaptic Mediator. Agent that mediates transmission at synapses between preganglionic and postganglionic sypaihetic endings *e.g.,* acetylcholine

Synachrony. Co-ordinated movement of particular part/parts of the body at the same time. *e.g.,* cilia of transverse row of a *Paramecium* beat in this manner In contrast to ,that of longitudinal row which have metachronous beating.

Syncitium. Undivided mass of cytoplasm containing many nuclei as found In the hypodermis of *Ascaris* sp. (nematode). It is a product of nuclear division without accompanying the cytoplasmic division or cytokinesis.

Syncitio Trophoblast. Blastocyst in contact with endometrium (innermost lining of uterus) becomes surrounded by an outer layer of multinucleate mass with no distinct cellular boundaries which is called syncition trophoblast.

Sydactily. Webbed fingers or toes as found in hind limbs of freg.

Syndrome. A group of symptoms and/or signs which occurring together produce a pattern or symptom complex typical of a part contraction of a particular disease.

Syneresis. Contraction of a Jelly-like mixture so that a liquid is squeezed out from molecular interstices.

Synergistic Muscle. A muscle co-operating with another one.

Syngamy. Fusion of gametes forming a zygote. Opposite to hologamy In which two individuals fuse throughout their body to foml a zygote.

Synovial Cavity. Cavity between two synovial membranes at articular surface between bones.

Synovial Fluid. Fluiq that is filled in the synovial cavity (capsule) forming a double walled structure binding particular caps. It contains large polysaccharides of hyaluronic acid and protein that give it a sticky and glairy consistency. It is secreted by synovial membranes and selves to lubricate particular surfaces of bones.

Synasacrum. Fusion of 3-4 sacral vertebrae as seen in case of mammals.

Synthesis. Joining of at least two molecules which results in formation of larger molecule.

Synteny. Association of genes on same chromosome. Often used synonymously with linkage. Some Genetists prefer to reserve the term linkage for only those genes which have been shown by genetic analysis to fall into a linkage group for example two genes shown by cell hybridisation to occur on the same chromosome would be considered syntenic. They would be called linked genes only if they do not assort independently.

Syronx. Voice box of birds, a hollow resonating chamber called tympanum inside which projects a semilunar membrane mounted over a cartilage pessulus. Vibrations set up in the membrane during the passage of air produce sound and the pitch is controlled by certain muscles present between syrinx and the body.

Systemic Arch. Large artery/arteries in a vertebrate corresponding to 4th embryonic aortic arches. It carries blood from heart to dorsal aorta. In fishes, amphibians and reptiles there are one pair of such arches but in birds and mammals there is a single systemic arch.

T

Tabes Disease. Disease of dorsal root of bladder. When sacral dorsal roots are interrupted all reflex contractions of bladder are abolished and the bladder becomes distended, thin walled and hypotonic.

Tachyauxesis. Rapid or fast growth of particular part of body as compared to whole body.

Tachycardia. Increase in heart rate above the normal level.

Tachyglossida. Family of monotreme mammals having large brain with convoluted cerebral hemispheres - spiny-ant-eaters.

Tachytelic. Species. which from fossil evidence, has evolved rapidly.

Tadpole. Aquatic tailed free-swimming larval stage of frog and uj)er animals like *Herdmania.* The larva has gills. segmented body. well developed alimentary canal for herbivorous diet, and after leading a free-swimming life for some period it metamurphoses into adult.

Taenia. Ribbon-shaped band of nerve fibres or muscle.

Taeniafuge. Any substance that can expel tapeworms.

Taeniodonta. Ordu of fossil quadruped terrestrial mammals that lived during Tertiary period in North America.

Taenioidea (Cyclophyllidea). Order of tapeworms (Platyhelminthes) that are mainly parasitic on warm blooded animals - *Diphlidium* species. *Hymenolepis* nana etc.

Tagma. Body region such as thorax or abdomen.

Tagmosis. Union of body segments into functional groups.

Tail. Postanal extension of body as found in all chordates. It may not be present in adult condition in few cases like frog. man etc. but certainly present in embryonic stage. A few non-chordates may have some structure looking like a tail but it is not a true tail since anus is situated at its tip.

Tail Diphycercal. Tall in a bony fish *Neoceratodus* (llung fish) is perfectly symmetrical and plirnitive because ventral lobe of fin is not supported by hypurals but by separate radials.

Tail Gephyrocercal. Tail in some fishes where tail fin of homocercal tail becomes secondarily simplified and tapering like diphycercal one.

Tail Heterocercal. Tail in fishes like *Scoliodon* has small upper lobe (epichordal lobe) and large lower lobe (hypochordal lobe) and thus it is asymmetrtcal externally as well as internally.

Tail Homocercal. Tail in teleosts like Labeo has equal hypochordal and epichordal lobe but end part of the tail (vertebral column) is bent upwards making it asymmetrical internally.

Tailor Bird Warbler. Small insectivorous. singing-bird, *Orthotomus* species (Passeriformes: Aves) of southern Asia and Australia that sews leaves together to form its nest.

Takin. Large unique mammal, *Budorcas* species which is half goat and half antelope, shoulder height about one metre, confined to Tibet, feeds on bamboo.

Talapoin. Small West African guenon monkey *Cercopithecus talapoin* (Arthropoidea: Mammalia).

Talbot. White hound (Carnivora: Mammalia) now extinct, predecessor of blood hound (extant dog); it had broad mouth with heavy jaws and large hanging ears; was used for hunting and tracking.

Talus. Tibiale or astragalous in mammals, the tarsal associated with tibia.

Tamandua. Two-toed ant-eater (Monotremata: Mammalia), no teeth. long slender tongue to lick up ants and termites; lives in trees grips branches of trees with prehensile tail, found in South and Central Arnerica.

Tamanoir. Giant ant-eater of South America, having remarkably long cylindrical snout and huge bushy tail.

Tamarin. Small squirrel-monkey found in South America. TANAGER: common name for brilliantly coloured birds of Tanagrid:e family (Passeriformes: Aves) found in forests of tropical America, feeding on fruits and insects.

Tapetum. Reflecting layer behind retina of many nocturnal animals, responsible for coloured reflection when a torch is flashed on such eyes - cat etc.

Tapeworms. Group of parasitic flatworms (Cestoda: Platyhelminthes), mouthparts and digestive system wanting. adult clings to intestinal wall of various vertebrate animals including human beings.

Tapir. Stoutly built mammal. *Taptrus tndtcus,* shoulder height one metre, snout prolonged into short trunk, body covered with short brtstly hair. found in wet tropical forests of South-east Asia and South America.

Tarantula. Wolf -spider *Lycosa tarantula* of South Europe (Italy), about 2.5 cm long, greyish in colour, with painful bite. Also large hairy bird-eating spider of South America.

Tardigrade. Class of Arthropoda that includes tiny, cylindrical, unsegmented, invertebrates with four pairs of stumpy unjointed legs - water-bears, found in damp moss and lichens. feeds on plant juice.

Target Cell. Cell that preferentially bind and respond to specific hormones.

Tarpon. Silver king: marine and fresh-water game-fish about 2.5 metres long, lives along tropical and sub-tropical coasts of Atlantic.

Tarsals. Bones found in ankle in the hind limb of vertebrates.

Tarsier. Small nocturnal primate (Mammalia): big eyes, large round sucker-like discs on fingers and toes, thick woolly fur: found clinging to tree trunk, feeding on insects. only three species, all confined to East Indies.

Tarsometatarsus. Bone of hind limb in birds formed by fusion of distal tarsals with the metatarsals in order to provide it strength to support the weight of body when the bird lands. In male it bears a fighting spur that helps in fighting dUring sexual selection.

Tarsus. Distal end of leg of an insect. It consists of one or more segments the distal most bearing claws. Also refers to ankle bone of vertebrates.,

Tasmanian Wolf. Dog-like carnivorous marsupial (Mammalia) about one metre long with tail another half a metre; light-brown fur with darker stripes across hind part of back; found in Tasmania and rarely in some parts of Australia.

Taste. Chemoreception in organisms for near objects enabling to discriminate between desirable and undersirable food.

Taste Buds. Receptors for taste in, form of minute barrel-shaped structures carried over small elevations in tongue called taste papillae. Two kinds of columnar cells constitute these structures - one which are centrally placed neurosensory cells and other that are peripheral non-sensory cells, project over the surface as in nammals or project into small pit-like taste pores opening on the surface. They are supplied by branches VII. IX and X cranimal nerves.

Taste Papillae. Elevations in tongue carrying taste buds. Four types of these are found in most vetebrates. (i) Filiform papillae - which are usually confined to anterior 2/3rd of tongue and rarely have taste buds. In carnivores (dog. cat. lion etc,) they are covered with sharp horny caps which help to pull out minute pieces of flesh from bones, (ii) Fungiform papillae - which are less numerous than former and have few-taste buds, (iii) Foliate papillae-which are leaf-like and found near the base of tongue on each side and bear taste buds on their lateral sides. (iv) Circumvallate papillae- which are largest papillae of rounded form having slightly raised central areas and low trench-like margin. Taste buds lie on lateral sides mostly in trench-like margin. Taste papillae are absent in whales..

Taurocholates. Sodium and potassium salts of taurocholic acid that are present In bile juice and help in preventing putrefaction of food within the alimentary canal.

Taurocholic Acid. Conjugates of acids (deoxyrholic acid and lithocholic acid) formed in the liver are glycholic acid and taurocholic acid.

Taxis. Oriehtational movement; may be towards a desirable or away from a undersirable condition.

Taxonomist. Person who has especialised in taxonomy.

Taxonomy. Science that deals with classification of organisms.

Telamon. Curved chitinous outgrowth of cloacal wall In various male nematodes.

Teleology. Scientific method of thinking. The use of purpose in which a result is treated as a cause in explanation of natural phenomenon.

Telophase. Any of the final phases of mitosis in which cells mayor may not divide but daughter nuclei are always formed

Teleoptiles. Final adult feathers in succession of feathers in a bird (compare with me sop tiles such as prefiloplumes, preplumes, prepinnae).

Telereceptors. Distant receptors which are concerned with events like distance. Chemoreceptors concerned with smell may be regarded telereceptors.

Telocentric. Chromosome with a centromere so close to one end that one arm is tiny or inconspicuous.

Telodendria. Axon ends in a number of synaptic buttons containing granules or vesicles in which' synaptic transmitter secreted by nerve is stored.

Telosporea. Class of Sporozoa (Porifera:Protlsta) having spores, without polar capsule. developing from oocyst.

Telson. Terminal extension of last abdominal appendage of a crustacean such as cray-fish, lobster etc.

Teleostel. Group of higher bony fishes including vast majority of modem liVing bony fishes like *Labeo, Clarias, Exocoetus, Anguilla* etc.

Teliolecithal. Mesolecithal egg: egg with large amount of yolk. as found in fishes and amphibians.

Template. Mould or pattern for the formation of a duplicate as in gene duplication.

Temporal. Lateral areas of head above the zygomatic arch.

Temporal Synapsm. Single fossa behind the orbit on each side and bounded by postorbital and squamosal above and jugal bones below. Seen In extinct reptiles and lower mammals.

Temporal Diasid. Characterlsed by two fossae - upper supratemporal and lower infraorbital. The former Is bounded by pareital and pre-frontal above and squamosal and postorbital below, while later Is bounded by squamosal' and postorbital above and quadratojugal and quadrate below. Found mainly in dinosaurs and crocodilians and Its modified forms are found in lizards and snakes.

Temporal Lobe. Lobe of cerebral hemlshere situated In lower lateral area at temples containing centres for speech and heartng.

Temporary Homeostatic Plug. When a blood vessel is damaged, the endothelium Is disrupted and an underlying layer of collagen is exposed which attracts platelets that adhere to it and liberate serotonin and ADP. ADP in turn attracts other platelets and a loose plug of aggregated platelets Is formed at the damaged part. Formation of this plug is unaffected by clinically used doses of heparin and dicumerol etc.

Tensors. Prepataglal muscles of birds namely tensor longus. tensor brevis, tensor accessorious and tensor posterius. Muscles that are useful for stretching any structure.

Tentacle. Flexible arm-like extension of body as found in many non-chordates like *Hydra. Obelia* etc. which helps in grasping and the locomotion. These are also used as offensive organs.

Tentculocyst. Sense organ found in some cnldartans like jelly-fishes. It is situated around the margin of umbrella and contains calcareous crystals arid sensory cells. It helps the animal to balance the body dUring locomotion.

Tergum. Notum: dorsal portion of exoskeleton of any body segment in animals like arthropods (cockroach etc.). It is joined to ventral piece called sternum with the help of lateral plates.

Terminal Cisternae. Portion of sarcoplasmic reticulum on both the sides of Z-line making close contact with T-tubules in a striped muscle fibre. (see striped/skeletal/voluntary muscles).

Terminating Codon. DNA base triplet which causes the cessation of protein synthesis; the terminating codon does not code for any amino acid *e.g.*, UGA

Terminology. Science of nomenclature: art of naming an organism or specimen.

Terrapins. Members of reptilian group Chelonia and close relatives of turtles. They are identified by characteristic angular rings on card-pace and are used as human food.

Territoriality. Behaviour of maintaining personal living space, used for breeding purposes. It may involve a bird chirping around perimeter of the territory or aggressive actions of animals whose territories are being violated.

Test. Shell over the body of many animals like *Herdmania* (where it is made up of tunicin), *Arcella* (pseudo chitin), *Polystomella* (calcareous) and also in a number of other protozoan animals belonging to group Testacea.

Testesterone. Hormone derived from testes and resposible for development of secondary sexual characters of male, like beard, hair on the body, harsh voice, aggressiveness etc., alongwith proper development of reproductive organs, the proper functioning of reproductivce organs in an individual. It also influences sexual behaviour and development of male accessory genetalia.

Testudinata. Equivalent name for Chelonia (Anapsida : Reptilia) that includes terrapin *(Trionyx gengeticusJ.* tortoise *(Ltssemys pWlCtata)* and turtle *(Chelone mydas)* etc.

Test Vesicle. Numerous tiny structures between test and mantle in ascidian tadpole which secrete the test.

Tetanus. Condition where a muscle undergoes permanent contraction and never relaxes again. This is due to the fact that the muscle is subjected to repeated stimulations during its contraction phase for a very long duration in case of experimental muscles. In man it is due to a pathogenic disorder caused by bacteria *Clostrtldium tetani* present in the dust and manure. Patient dt elops spasm and rigidity of muscle and his body becomes stretched into bow-shape leading finally death. AIS injections produce passive immunity for the disease.

Tetany. Condition of muscular hyperexcitability in which mild stimuli produce cramps and spasm.. In infants it is associated with gastrointestinal upset and rickets.

Tetany Hypocalcemic. Trousseaus Sign: a disease due to increased activity of motor nerve fibres and characterised by extensive spasms of skeletal muscles of exteremities and larynx: caused due to fall in Ca++ions of blood due to disturbed calium regulation by parathyroid.

Tetra Corallia. Old name of Rugosa. an order of extinct corals (Coelenterata).

Tetradactylous. (animal) haVing four fingers or toes on each limb.

Tetranychidae. Family of Acarina (Arachnida: Arthropoda) that includes spides-mites; phytophagous; spin silken webs on leaves and other plant parts and feed under the protection of these webs. sucking cell sap. causing rusty discolouration spots - *Ollgonychus mangiferus* on mango, grapes. jamun etc.. *Tetranychus ctnnabarinus* on almost all fruit trees and vegetables.

Tetrapooda. Group of vertebrates having four limbs *i.e.,* one pair of anterior or fore limbs and a pair of posterior or hind limbs. The group includes all the vertebrates above the level of fishes i.e., amphibia. reptiles. birds and mammals.

Thalamic Nuclei. Nuclei present' in the thalamus and its parts (epithalamus, dorsal thalamus and ventral thalamus).

Thalamus. Portion of vertebrate fore brain; consists of epithalamus. dorsal thalamus and ventral thalamus.

Thalassemia. Inherited blood disorder in which homozygote suffers with a severe anaemia called *Thalassemia* major and the heterozygote suffers with milder form of anaemia called *Thalassemia minor.* It is actually a general term which includes various anaemias due to effects of different genes.

Thebesian Valve. Tricuspid valve which is present at right auriculoventricular opening of mammalian heart. It controls back-flow of blood from right ventricle to right auricle during ventricular systole.

Thecodont Teeth. Teeth that are lodged into definite the co or cup-like structures formed within jaw bones. These teeth are found exclusively in mammals. Compare with acrodont and pleurodont teeth of other vertebrates.

Theroy. Possible explanation of natu.al phenomenon for which some evidence exists but no definite proof is available.

Thermophile. Organism that thrives at high temperature.

Thermoreceptors. Receptors in the body of organisms which respond to temperature of the surrounding. They are present on outer surface of body in the form of nerve endings. specialised structures like ampullae of Lorenzini (found in cartUagenous fishes like *Scoliodon)* or'some other form.

Thermoregulation. Ability of an organism to regulate its body temperature. It includes morphological. physiological or behavioural adaptations within limits necessary for its survival. Human beings have constant body temperature of *3TC.* an increase in environmental temperature above this temperature results in sweating by the body which brings about a heat loss and similarly a fall in environmental temperature below this causes shivering etc. so that the body's heat loss is checked.

Thermotaxis. Behavioural response of an organism to heat. If *Euglenae* are kept in a tube and one end of tube is kept warm (above 40' C) and the other kept cold. they do not remain either towards the wann end or cold end but are seen to move and settle midway where they get optimum temperature.

Thesocytes. Cells found in the mesoglea of sponges like *Sycon* which have amoeboid shape and a large nucleus. There are several rod like inclusions in the cytoplasm which are possibly reserve food materials: function to store reserve food like liver in higher vertebrates.

Thetarhythm. Pattern or regular waves in electroencephalogram where number of waves is 4-7/second This rhythm is seen generally in case of children. A rhythm having less than 4 waves/second is called delta-rhythm and more than this (18- 30/second) forms β- rhythm.

Thin Filament. Filament in striped muscle which is composed of another protein called actin.

Thiouracil. Compound related to thiourea and known to inhibit inodination of monoidodotyrosine and thus blocks coupling reaction and inhibits thyroid function.

Thirst. Feeling to take fluid. It is generally stimulated by decrease in extra cellular fluid volume and processes like haemorrhage etc. It is under hypothalanic control (see thirst centre).

Thirst Centre. Sunfornical organ in diencephalon which acts like a centre for stimulating thirst.

Thoracic Pump. During inspiration intrapleural pressure falls from -2.5 mm to -6 mm of mercury. This negative pressure is transmitted to great veins and to lesser extent to atria so that central venous pressure fluctuates from about 6 mm of mercury. dUring expiration. to approximately 2 mmof mercury dUring quiet inspiration. The drop in venous pressure during inspiration aids venous return. When diaphram descends during inspiration. intrabdominal pressure rises and this also squeezes blood towards heart because back-flow into large veins is prevented by venous valves.

Thorax. Major division of vertebrate animal just posterior to neck or the head if the neck is absent: part of body of arthropod situated between head and abdomen.

Throxoxane. Most active cell regulaters fonned from prostaglandin precursors. Two types - thromboxane-A2 and thromboxane-

B2 (both being fonned from endoperoxldase).

Thylacine. Tasmanian wolf: flesh-eating marsupial *Thylactnus cynocephalus* (Carnivora: Mammalia), greyish-brown with conspicuous black markings on posterier half of back.

Thymosin. Polypeptide extracted from thymic tissue and thymic honnone that promotes maturation of T-lymphocytes in young condition and maintains immunity reactions during that period. The substance contains 108 amino acid residues and is rich in acidic amino acids.

Thymus. Two lobed endocrine gland located Just below thyroid of *mammals.* They are developed during childhoOd when they play defensive role in the body by secreting thymic honnone which promotes maturation of T-lymphocytes. In puberty they disappear because their secretion hampers the development of reproductive function.

Thyrocalcitonin. Calcitonin is often referred to as thyro-calcitonin.

Thyroglossal Duct. Duct marking the path of thyroid from tongue to neck (often persisting in the adulthood).

Thyroglobulin. Globulin that binds tyrosine molecules in peptide linkage(from tyrosine are synthesised colloids by iodination and condensation). This glycoprotein has 4 peptide chain and molecular wight of approximately 670000. It is synthesised by thyroid cells and secreted in the colloids. Honnone remains bound until secreted. After secretion colloid is ingested and honnones are released into capillaries.

Thyrohyal. Posterior horns of hyoid apparatus representing fourth branchials.

Thyroid Cartilage. Pair of cartUages In the larynx of mammals which enlarge to fonn a protuberance in the neck region. called Adam's apple. in adult man. Vocal cords extend between these cartilages and arytenoids.

Thyroid Glands. Paired glands in the neck region of vertebrates which are endocrine in nature. They secrete thyroxine honnone which regulates body growth. Their activity Is controlled by thyroid stimulating honnone from anterior pituitary lobe. Hyposecretion by these glands leads to myxodema and hypersecretion leads to cretinism.

Tidmann's Bodies. Spherical swellings (caeca or pouches) on the inner wall 'of ring canal of water vascular system of echinoderms like star-fish *(Pentaceros* species); generally nine in number; these have gla'ndular epithelium and are supposed to be, concerned with oroduction of amoebocytes.

Tigers. Large, fierce, flesh-eating beasts *Panthera (Felis) ttgris* (Carnivora: Mammalia), having black striped tawny coat with whitish belly; normal length little over 3 metres (including tail) though some reach almost 5 metres; found in almost all countries from Indonesia to Iran in mountainous regions and dense forests; fast runner and good swimmer but do not climb well, some species such as Balinese, Caspian and Javan etc., are extinct by now.

Tiger Beetles. Small brightly coloured beetles (Cicindellidce: Coleoptera); both grubs and adults feed on other insects.

Tiger Moths. Medium- sized, stout-bodied, brilliantly coloured moths (Arctiidae: Lepidoptera), dark patches on white forewings and red or orange hind wings; caterpillars densely clothed with hair.

Tiger Shark. Huge voracious shark. *Stegostoma* species having brilliant dark brown and white stripes; grows up to 5 metres in length; inhabits Indian ocean.

Tiger Snake. Highly poisonous Australian snake, *Notechis scutatus* (Ophidia: Reptilia) having brown body with black cross-bands.

Tiger Wolf. Spotted hyaena (Carnivora: Mammalia).

Tight Junction. Cell Junction in which membranes of two cells become apposed and the outer layers of membranes fuse with each other.

Tigon. Offspring of tiger and lioness (Fissipedia: Carnivora).

Tinamiformes. Monofamilial order or retite birds (Palaeognathae Neornithae). fowl-like birds but haVing fully developed (junctional) Wings; weak fliers; found in Central and South America.

Tissue. Group of cells similar In structure (with Intercellular substances. If any) which perform a spec1alised functlon. The study of tissues Is known as histology. The main types of tissues found In the body of a vertebrate are epithelial tissue. which froms the various coverings In the body. connective tissue which establishes connection between various body parts by forming blood. muscles. bones etc. and the nervous tissue that forms a co-ordlnating system among various body structures and controls the\r functioning.

Tissue Macrophage. Monocytes enter circulation from bone marrow but after 24 hours they enter the tissues to become tissue macrophages. All kinds of tissue macrophages Including Kupffer's cells of liver. alveolar rnacrophages of lungs come from circulating monocytes.

TITI. small south American monkey *Callithrtx* species (Platyrrhlna Mammalia).

TIT Mouse. Chickadee or tit : group of small. acrobatic song birds of northern hemisphere - blue -tit. *PantS caeruleus;* coal-tit. *Pater:* great-tit. P. *major* etc.

Toad. Leaping tailless amphibian (Aglossa: Annra) with no tongue and no teeth-Pipa *americana, Bufo melanosttctus. Bombinator tgntius* etc.

Tocology. Obstetrics: science concerning childbirth. TOCUS childbirth.

Tone. Tonus: normal healthy state of tension in skeletal muscle.

Tongue. Mobile. muscular organ contained in the mouth cavity. It is concerned with speech. mastication. swallowing and taste. It is mostly joined posteriorly and free anteriorly but In certain cases like frog it Is jolIl,td anteriorly and free -posteriorly. In snakes and lizards it is bifid.

Tozicology. Scientific study of poisonous substances and their effect on different parts of organisms.

Toxin. Product of bacteria. poisonous to other cells which on injection into animal or a man stimulates the production of antibody to it. Non-bacterial toxins refer to poisonous nitrogenous compound such as liberated fmm diseased or injured tissue.

TPN (*Tri-phosphopyridine Nucleotide*). Coenzyme-II or NADP. one of the substances that pick up hydrogen ions formed during dehydrogenation in biological oxidations.

TpT. tropmyosin binding subunit of troponin. It is found in association with actin of thin filament.

Trachea. Wind pipe in vertebrates; the fibrocartllagenous tube lined with mucous membrane passing between larynx and bronchi. Its lumen bears ciliated cells and mucous producing cells, the connective tissue supporting the walls is supplied with blood vessels. The structure is also found in insects where it is connected to outside through splracles. It divides into several branches which are called bronchi, bronchioles etc. which form respiratory tree in vertebrates.

Tractus Solitarius. Myelinated but slow-conducting taste fibres in VII. IX. and X- cranial nerves on each side unite in medulla oblongata to form tractus solitarius.

Tragus. Small prominence in front of opeping of external ear (hairy in man).

Trail Endings. Nerve endings of gamma efferent fibres that form extensive network primarily on nuclear chain ,fibres.

Translocation. Chromosome aberration in which a segment of chromosome is shifted to a new location in the same or completely different chromosome.

Transphosphorylation. Transfer of phosphate group between organic molecules by passing the inorganic phosphate stage.

Transexual. Individual who considers himself/herself to be person of either sex. It is not considered a homosexual since the later person identifies with his or her anatomical sex

Transeverse Process. One of the lateral projections in a vertebra of the vertebrates.

Trauma. Bodily injury and also emotional shock.

Traumatic Shock. Shock resulting. to body from an injury or emotional disturbances.

Tree Duck. Whistling duck : large goose-like bird (Anseriformes :Aves); nests on ground near water; eight species distributed throughout tropical and subtropical regions.

Tree Frog. Arboreal frog *Hyla arborea* (Hylidre: Anura) and allied species, climb and cling to surface of leaves; generally green in colour, *some* species are capable of changing colour according to their background to avoid detection by their predators.

Tree Kangaroo. Tree-dwelling marsupial, *Dendrolagus* species (Metatherl a: Mammalia) of Australasia. about 75 em long body and furry tall another 75 em.

Tree Porcupine. Arboreal porcupines of New World; some species of South America have prehensile tail.

Tree Shrew. Small sqUirrel-like insectivore mammal *Tuparia* species (Tupaiidre, generally brown in colour and about 15 to. 20 cm long body with tail another 10 to 20 em.

Treppe. Staircase phenomenon: condition during muscle con taction as seen In kymogram, when a muscle is stimulated repeatedly after regular 'intervals: for some time. the initial few stimulations show gradual increase in amplitude of contraction but successive later ones show a decreasing amplitude.

Triads. Arrangement of central T-tubule with sarcoplasmic reticulm of skeletal muscle fibre on either side which functions to help in rapid transmission of action potential from cell membrane to all the fibres in muscle.

Triceratops. Gigantic three horned dinosaur that roamed the earth during Mesozoic era.

Trichoptera. Order of moth-like insects (Hexapoda) having soft body, setaceous antennae rudimentary mouthaparts, membranous wings that are densely hairy, metomorphosis complete – caddice flies.

Trichocyst. One of the definite bodies lying In the ectoplasm and producing hair-like fibres in organisms like *Paramecium.* These structures are discharged on chemical. mechanical stimulation as well as at the time of death of the animal.

Trigeminal. Fifth cranial nerve in vertebrates which divides into three branches supplying the skin of face. tongue and the teeth.

Triglycerides. Esters of glycerol and 3 fatty acids. They yield glycerol and 3 fatty acids on being hydrolysed *e.g.,* deshi ghee (a natural fat). Formed in animals from carbohydrates.

Trisomy. Chromosomal mutation leading to an individual possessing hee copies instead ofnoral twocopies (diploidy) of a given chromosome.

t-RNA. S-RNA (souable-RNA). small RNA molecule which is able to recognise a specific amino-acid, transport it to ribosome and position it properly in growing polypeptide chain.

Tropism. Antiquated term for orientation and locomotory responses *e.g.*, phototropism means orientation and locomotory responses with respect to light similarly heliotropism pertains such movements with respect to Sun. Other such tropic movements are geotropism (with respect to gravity), thigmotropism (with respect to touch), hydrotropism (with respect to water) etc.

Tropomyosin. Long protein molecule located in the grooves of actin of thin filaments of strtped muscles. It inhibits contraction of myosin cross-bridges with actin filaments.

Troponin. Complex of globular protein associated with actin and tropomyosin of thin (actin filament) of strtped muscle. It has three subunits:

Truncus Arteriosus. Base of aorta as found in lower vertebrates like frog. In the lumen of this is found a spiral valve which serves to protect the collapsing of its walls as well as distribution of blood through the organ.

Trypsin. Enzyme promoting digestion of proteins in the vertebrates. Found in the pancreatic juice of all the vertebrates.

Trypsinogen. Inactive form of trypsin in which it is secreted in order to prevent autodigestion. The substance is activated by enterokinase from succus entericus.

T-System. One of the structural elements of sarcotubular system that consists of transverse tubules which is continuous with membrane of mu.scle fibre. The space between two layers is an . extension of extracellular space.

T-Tubules. Running around the perimeter of each myofibril at the level of Z-line there is membrane limited tubule called T-tubule (transverse tubule). It is less than O.Iμm in diameter and branches so that it is continuous with similar tubules surrounding neighbouring myofibrils at the same level.

TTX (*Tetrodoxin*). Puffer-fish that selectively blocks Na+ ion channels in excitable cell membranes.

Tube Feet. Locomotory organs found in the ambulacral grooves of star-fish and other echinoderms. Each is connected with water vascular system and consists of dorsal sac-like ampulla and ventral disc-like sucker. Certain muscles over the ampulla and the sucker regulate contraction and expansion of these parts and in turn help the organ to attach with the substratum or relax the hold of substratum. It is by bringing about such necessary contractions and expansions. the animal performs locomotion with these organs.

Tubular Fundimular System. Separate intrahypothalmic system of dopaminergic neurons projecting from cell bodies into accurate nucleus to external layer of median eminence of hypothalrnus.

Tubule. Small tube-like structure present in the body. It may be : Coneeting tubule: associated with kidney and conveying urine to ureter. *Convoluted tubule*: present in anterior part of renal tubule. *Uriniferous tubule*: tubule syonyrnous with nephron. *Seminiferous tubule* : tubule associated with testes.

Tubulin. Actin like 40 A globular protein molecule that is building block of microtubule.

Tufted Cells. Cells among dendrites ofaxons of receptors which terminate in olfactory bulbs. Axons of tufted cells pass posteriorly through the medial olfactory stria. cross the middlel1ne in anterior commissure and enter the opposite olfactory bulb.

Tumour. A swelling. a mass of abnormal tissue which resembles the normal tissue in structure but fulfils no useful function and grows at expense of body. Benign/simple/innocent turnouTS arc encapsulated and do not infiltrate adjacent tissue. unlikely to recover if removed. Malignant tumours are uncapsulated and infiltrate adjacent tissue and cause metastasis.

Tunica Media. Middle muscular coat of artery.

Tunica Vaginals. Peritoneum of vaginal coelom reflected over a part of testis after its descent into sacrotal sac in case of mammals. Increase in fluid within this layer causes hydrocoel.

Tunicate. Group of animals with test over the body - *Herdmanla* species (Urochordata) .

Turbinals. Thin scroll-like folded bones in the olfactory part of nasal chamber of mammals. They are covered with an epithelium which is highly vascular and contains olfactory receptor cells, mucous cells, supporting cells etc. The presence of these structures helps the olfactory chamber to maintain constant temperature of the air that goes into the respiratory tract.

Turbulence. Phenomenon in blood characterised by disorderly flow occurring when viscosity of blood is lower and is related to the diameter of vessel. Turbulent flow also creates sound while . streamline flow is silent.

Turbulent Flow. Flow in which fluid moves in all the directions. It causes sound as opposit~ to sUent streamline flow.

Turner's Syndrome. Gonadial dysgenesis that can be identified at birth. Individual of this syndrome has 45 chromosomes instead of 46 and one sex chromosome is missing. (XO sex chromosome constitution). Brought up as girls though they are genetically male. Such an individual has small female genetalia, scanty pubic hair, atrophic ovaries, webbed neck and valgus of elbow. Usually sterile. The syndrome was discovered by American endocrinologist Henry H. Turner.

Turtle. Semi-aquatic carnivorous, marine reptile (Chelonia: Anapsida), soft-body encased in hard shell (carapace), fore-limbs modified as flippers fot swimming.

T-Wwave. Portion of electrocardiogram associated with repolarisation of ventricles (usually relaxation).

Twins. One of the two individuals born at same birth: one very much alike to another.

Twin Fraternal. Dizygotic twins that develop from two eggs ovulated at the same time; each zygote passes separately from Fallopian tubes to uterus; each embryo implants separately. develops its own extra-embryonic membranes. is nourished by its own placenta and floats in its own amniotic cavity; the twins may be of same or different sex and have always different genotypes.

Twin Identical. Monozygotic twins. result when a single embryo splits during its development into two embryos usually within a fortnight of fertiHsation. paired foetuses haVing same extra-embryonic membranes and a single amniotic cavity; two individuals are of same sex and genotype.

Twin Siuamese. Rare phenomenon. when space available for development of monozygotic twins is limited, the paired foetus. occupying a single amnion, get fused, resulting in conjointed twins.

Twitch Muscles. Most common striated vertebrate skeletal muscle type usually pale in colour because of low myoglobin contents. It has few mitochondria and its fibres are constructed of many clearly defined fibrils. 'These contract rapidly and derive most of energy through anaerobic metabolism.

Tylopoda. Group of ungulate animals having padded toes - camels, Ilarnas.

Tympanic Bone. Bone derived from angular bone of reptilian lower jaw (in mammals which is closely applied to petrous part oi periotic bone).

Tympanic Bulla. TympanIc bone of mammals is closely applied to a part of periotic bone and forms flask-shaped structure called tympanic bulla; the neck of this structure forms external auditory meatus.

Tympanic Cavity. Air filled space/cavity in the middle ear comIl1unicating with pharynx by means of eustachian tube.

Tympanic Reflex. Sound transmission reduction by contractionof tensor tympuni and srapedial muscles pulling manubrium of malleus inward and foot plate of stapes outwards, initiated by loud sound.

Typhlosole. Median dorsal intestinal fold of mucous membrane in the intestine of earthworm; serves to increase area of absorption for ntestine of the animal.

Tyrannosaur. Flesh-eating dinosaur (Reptilia) of Cretaceous period (now extinct); forelegs extremely small and non-functional; walked only on hind legs, balancingits huge body with the help of long powerful tail.

U

UDAD. Barbary sheep: large-horned. long-tailed wild sheep of Africa having a fringe of long hall' hanging from throat and chest (Aruodactoyla : Mammalia).

UDP GGlucuronyl Deficiency Type-II. Congenital disease in which there is relative deficiency of glucuronyl transferase that catalyses conjugation of free bilirubin to glucuronic acid.

Ugly-best Caterpillar. Arclps ceraslvorana (Tortrlcldre Lepidoptera) are commonly found In tropical regions, on black cherry and choke cherry trees. They live in colonies and enlarge their nests by covering the adjacent foliage that Is used as food; such nests give an ugly appearance.

ULNA: The bone on little-finger side offore arm or fore limb. found in vertebrates - from amphibians to human beings.

Ultrafibril. Smallest unit of chromosome visible under an electron microscope. It is 100 A thick and consists of two DNA double hellces, separated by space of 20 A across the associated protein. Smallest ultrafibril Is single DNA with its associated protein. As single DNA molecule has roughly 20 A diameter. two DNA molecules will have 40 A thickness and a gap of 20 A between two DNA helices of ultrafibril will have a total thickness oflOOA.

Ultrafiltraton. Within glomerulus when the blood passes from capillaries of efferent renal arteriole to efferent renal arteriole then because of pressure difference certain substances from blood are fallen in Bowman's capsule and then diffuse to the neck of renal tubule through one celled thick wall of capsule. This filtration Is called ultrafiltration.

UMBO. Elevation near the hinge of bivalve shell as seen in *Unto* species around which the concentric lines of growth are found In the shell.

Umbraculum. Specially modified part of iris in camels and other mammals in form of pigmented fringe selVing as umbrella to shield retina from excessive glare.

Umbrella - Ant. Sauba or parasol ant. *Atta* species (Hymenaptera: Hexapoda): leaf-cutting ant that carries pieces of leaves over its body, like an umbrella to the underground nest.

Umbrella Bird. Fruit - crow; medium-sized blue bird having large extensible head-crest resembling half-open umbrella, able to cover head and beak completely; found in South and Central America

Uncinate Process. Flattened median outgrowth from ribs in the birds

Undulipodium. Whip-like apparatus with 9+2 arrangement of inner ffiaments. It is a new term for cilium and flagellum.

Unguis. Long plate of keratin found in claws, nails etc., which is sharply curved so as to enclose always smaller sole plate or subunguts between its edges.

Unguiligrade. Walking or an adaptation for walking on hooves as seen in horse, sheep, cow etc.

Unicorn. One-horned: fabulous animal mentioned in Greek mythology that had body of horse and a single horn projecting from the forehead.

Unicorn Fish. Tropical fish (Pisces) with flattened body, numerous tiny scales and a horn-like outgrowth from its head.

Unicorn Moth. Lepidopterous insect whose caterpillar has a long horn-like process or appendage.

Unio. Fresh-water pearl-mussel (Mallusca), delicate,. bilaterally symmetrical body covered hy hard bivalve shell.

Unipotent Committed Stem Cells. Cells in bone marrow which on stimulation are differentiated into one of the cell types found in red bone marrow (monoblast. myeloblast. lymphoblast. proerythroblast and megakaryoblast).

Unisexual. Having only one sex. male or female.

Unit Membrane. Sandwich-like profile of biological membrane seen in electron microscope and bellieved to represent binocular leaflet with hydrophobic central region between hyrophilic surface.

Universal Donors. Type '0' indivduals which have no agglutinogen and thus cause agglutination when their blood is transfused to an individual belonging to any other blood group.

Uracil. RNA base which is analogous to thymine in DNA

Uremia. Clinical syndrome due to renal failure resulting from either disease of kidneys or disorders elsewhere in the body which induces kidney dysfunction resulting in retention of urea and other nitrogenous substances in blood. Fully developed syndrome is characterised by nausea, vomiting, headache. hicough. weakness, dimness of vision, convulsions and coma.

Ureotelic Animal. Animal that excretes urea in the urine-amphibians and eutherian mammals Cf. uricotelic.

Urethra. Duct that carries mine from bladder to outside in mammals.

Uric Acid. Crystalllne waste product of nitrogenous metabolism found in faeces and urine of reptlles, birds and egg-laying mammals. It is insoluble in water and can be safely stored In the body without causing any harm to the body of individual.

Uricotelic Animal. Animal that excretes uric acid in urine -reptiles. birds and egg-laying mammals.

Uridine Diphospho-Galactose. Substance fonned by reaction of galactose-6-phosphate with uridlne diphosphoglucose.This Is converted into uridine dlphosphoglucose

Uridine Diphospho-Glucose. Substance to which galactose (after being subjected to phosphorylation) reacts dUring glycogen formation from galactose.

Urinary Bladder. Thin sac-like structure attached with cloaca in which urine fonned by kidneys is accumulated.

Urine. Liquid waste excreted by kidneys in vertebrates.

Uriniferous Tuble. One of the excretOJY tubules within kidneys of higher animals consisting of a coiled lube and a capsule.

Urinogenital System. In vertebrates excretory organs and reproductive organs are closely associated since early development thus both systems are represented together as urinogenital system.

URO. Posterior part: tail portion.

Urochrome. Pigment that gives yellow cohur to urine.

Urodaeum. Chamber/division of cloaca in reptiles and birds in which urinogenital organs open. It opens into proctodaeum.

Urodela. Order of talled amphibians (Chordata) distinguished from Apoda by having legs and from Anura by presence of tail throughout their llfe - newts. salamanders.

Urogastrone. A substance that has inhibitory effect on stomach secretion. found in urine.

Urohyal. One of the median bones in hyoid apparatus representing the basibranchial of first branchial arch.

Urologist. Person well-versed in urology.

Urology. Scientific study of urine.

Uromose. Abdominal segment of arthropods.

Uterine Cycle. Cyclical changes in the structure and- function of uterus with respect to the hormones associated with menstrual cycle, shedding of endromltr1um, development of uterine glands, secretory or progestational changes, formation of stratum functionale and preparation of next sloughing in case pregnancy has failed.

Uterine Tube. Fallopian tube.

Uterus. Enlarged portion of an oviduct in which atleast part of development of the animal takes place. Technically the term is applicable mostly to an animal (like eutherian mammal) in which embryo becomes attached to the wall of the organ.

Uterus Masculina. Seminal vesicle.

Uvula. Soft pallate in mammals that ends in front of gullet as a median fleshy elongation, used to press posterior nares during ingestion.

Uropepsinogen. Pepsinogen activity detected in plasma and urine.

Uropod. Flattened abdominal appendage in crustacean like cray-fish.

Uropygial Gland. Preen-gland of birds placed at the end of tail.

Uropygium. Small conical tail-like projection in birds; bears preen/uropygeal glands on its dorsal side.

Urosthenic. Having tail developed for propulsion as in crocodiles.

Urostyle. Terminal rod-like bone in the vertebral column of frog.

Ursida. Family of carnivore mammals, plantigrade galt, elongated skull, head produced into snout, ears large, limbs thick and clumsy: feeds on fruits, vegetation as also goats, sheep etc.

V

Vacuole. Small structure enclosing a cavity in cytoplasm: may be filled with a fluid/liquid as in case of contractile vacuole or may contain other substances like food (food vacuole). secretory products (secretory vacuole) etc.

Vagal. Pertaining to vagus nerve (X-cranial nerve).

Vagina. Posterior part of female reproductive tract following usually the uterus (as in case of mammals), often receives copulatory organ of male during process of mating.

Vaginal Cycle. Cyclic changes in vaginal wall with respect to different hormones of reproductive cycle. These changes particularly in vaginal wall are cornification of vaginal wall's epithelium under influence of estrogen and secretion of thick mucous and proliferation of epithelium which becomes infietrated with WBCs under Influence of progesterone. The above mentioned changes are seen in expertmental rats and also reported to be occurrtng in human beings but not so clearly marked out.

Vaginal Smear. Film of material spread out on a glass slide for examination under microscope. The cells in vaginal lumen present in smear show conformational changes which help in detection of complication of vaginal cycle and show infection etc. which might be present in vaginal wall.

Vagotomy. Surgical division of vagus nerve done in connection with gastroenterostomy in the treatment of peptic ulcers or pyoloplasty.

Valine. Neutral amino-acid: essential amino acid with unsubsututed chains.

Valve. Fold of membrane in a passage or tube. permitting flow of contents in desired direction only and preventing back-flow. The valves are found at the Junctions between dIfferent divisions of alimentary canal and in the veins.

Vane. Distal portion of contour feather looking as flattened semitransparent structure on both the sides of rachis and composed of barbs. barbules and barbiceles (which interlock together).

Vas Efferens. Small ducts carrying sperms from testes to kidney tubules in some animals like frog and to epididymis in case of higher animals.

Vas Vasorium. Minute nutrient vessel of artel)' and vein walls.

Vascular. Pertaining to vessels in animals. these usually cany blood or lymph. The term is used mostly for a structure in the body having a vel)' rich supply of blood vessels and capillaries.

Vasectomy. Surgical removal of portion of each vas efferens thus blocking addition of sperms to seminal fluid.

Vasoconstrictor. Substance that induces constriction of blood vessels. Local vasoconstrlction is producted by serotonin and certain substances produced by some body structures. Renal vasoconstriction is produced by renin-angiotensin system.

Vasomotor Centre. Centre in medulla oblongata that controls discharge rate in sympathetic vasconstrictor nerves. A large diffused area in reticular formation from Just below the obex to region of vestibular nuclei and floor of fourth ventrtcle to almost pyramids. Stimulation of rostral and lateral portions of this centre causes rtse in blood pressure and tachycardia. While stimulation of a smaller portion around the obex produces a fall in the blood pressure (bradycarta).

Vasomotor Verves. Nerves that control contraction and expansion of blood vessels.

Vector. Carrier of diease *e.g.*, housefly, mosquito etc. It also refers to secondary host in the life-cycle of a parasite as mosquito in the life-cycle of *Plasmodium* and pig in the life cycle of *Taenia solium*

Vegetative Hemisphere. Part of egg where the rate of metabolism is lower than that of the animal pole; usually contains yolk material.

Veins. Blood vessels having flow of blood from postertor backwards to antertor forward direction. They generally carry deoxygenated blood and also have valves at places.

Velarium. Velum of certain cubomedusans and scyphozoans charactertsed by presence of canals lined with endoderm.

Verliger. Larval stage after trochophore in many molluscs like *Pila.* VELUM : membranous band between stomodaeum and pharynx of *Amphioxus.* beartng 12 backwardly directed tentacles along postertor border. Also called enterostome.

Venae Cavae. (*Caval Veins*) three large veins collecting deoxygenated blood from various parts of body and drain it into sinus venosus/right auricle. They are also termed superior and inferior venae cavae or pre-and post -cavals.

Ventral. Side/surface of body pertaining to belly; away from back. opposite to dorsal.

Ventral Horn. Ventral portion of gray matter in vertebrate spinal cord in which mother nerve cell bodies are situated.

Ventral Root. Nerve trunk leaving spinal cord near its ventral surface containing only motor neurons.

Ventricle. Any of the small chambers in the anatomy of animals;

term is strictly used with relation to chamber in the heart which is the chamber from which blood is distributed. In brain the term is used for several large divisions of central space.

Ventricular Arrhytmias. Deviation from normal rhythm in ventricle. This may be due to development of ectopic focus in ventricle that causes extrasystole/premature beat which may distrub the normal cardiao rhythm.

Ventridine. Drug that causes apnoea. hyptotension and bradycardia in experimental animals when injected into their body.

Vermis. Median lobe of large sized cerebellum in fishes and mammals.

Verrucomorpha. Suborder to Thoracica (Crustacea) that includes sessile asymmetrical barnacles.

Vertebrae. Structural unit of vertebral columns of a vertebrate. Each of them principally has a centrum(formed by replacement of notochord) and dorsal, ventral processes (that form neural spine), transverse processes, zygapophyses and hypocentrum.

Vertebral Column. Replacement of notochord by a bony structure consisting of vertebrae surrounding the spinal cord in all the vertebrates *e.g.,* fishes, amphibians, reptiles, birds and mammals. It may be differentiated into cervical, thoracic, lumber, sacral and caudal regions.

Vertebrarterial Canals. Pair of canals placed on either side of cervical (neck) vertebrre of birds and mammals especially, through which vertebral arteries pass.

Vertigo. Sensation of rotation in absence of actual rotation.

Vervet. Small greyish-green long-tailed African monkey. *Cercopithecus aethiops pygetythes* (Arthropoidea: Mammalia).

Vespidat. Family of Vespoidea (Hymenoptera: Hexapoda) that includes wasps like hornets, yellow jackets etc.

Vestigial Organs. The organs once tn use but now reduced, small and non-functional owtng to disuse *e.g.*, nictitating membrane in human eye (plica semilunaris), tail tn man (coccyx), appendix in the tntestine of man etc.

Villikinin. General name for different hormones secreted from gastrointestinal tract like VIP, GIP, bombesin, somatostatin etc, which regulate intestinal activity.

Visceral Sensory. Nerve fibres in the lower part of dorsal horn dealing with internal reflexes.

Vviscocity. Physical property of flUids that determines the case with which layers of fluid move fast to each other.

Visual Agnosia. Inability to see objects by a particular visual modality even though visual modality is intact. It Is due to leislons in parietal lobe.

Visual Threshold. Minimal amount of light that elicits a sensation of light.

Vitalism. Theory that biological processes can not adequately be explained by physical and chemical processes and laws.

Vitamins. Any of the unrelated organic substances that do not yield energy but their presence is must in diet. They are essential for normal growth and function and necessary in minutest quantities, Two catagories of these substances are. water soluble vitamins and fat soluble vitamins. Vitamin A, D. E and K are fat soluable while vitamin B. B-complex and C are water soluable.

Vitelline Membrane. Thin membrane surrounding the egg/yolk.

Vitorous Humor. Liquid within eye in the area between lens and retina. It is jelly-like in appearance.

Viviparity. Ability by animals to give birth to living young one that develops from egg within the body of mother and nourished with the help of a placenta.

Viviparous. Organism that gives birth to living young one that developed In the body of mother and nourished by her own blood stream as most of eutherian mammals.

Voval Cords. Two membranous folds extending from side wall of larynx:. repsonsible for pitch of voIce.

Voles. Group of mouse-like small rodents. having rounded snout. short ears and tail; herbivorous -' bank vole. *Clethrtonomys glareolus:* field vole, *Microtus agrestts,* water vole *Arvtcole amphtbius.*

Volkmann's Canals. Canals connecting Haversian canals in the mammals and found below periosteum and in the centre of bone. Through these canals blood vessel reach interior bone.

Volutidat. Family of Neogastropoda (Prosobranchia:Mollusca) that includes gastropodes like *Conus* species, *Murex* species, *Terebra* species etc.

Volvent. Small type of cnidarian nematocyst ccntaining an armed thread (with spines) used to capture prey.

Vomit. Spew: eject forcibly stomach contents through mouth.

Vomiting. Reflex action of ejecting stomach contents through mouth. Also considered a sickness due to deficiency of viamin or may be due to infection or other disorders of gastrointestinal tract.

Vortical Sounds. During turbulent flow if the blood strikes an obstruction or passes through a narrow orifice at high speed swiling vorticals set up generate vortical sounds.

Vultures. Group of large raptorial birds (Falconiformes: Aves), with head and neck almost naked; strong, curve bill, powerful feet and laws; fee mainly on carrion–ding vulture, sarcogyps caluus, white scavenger vulture, Neophron perenopterus.

Vulva. External Portion of femal reproductive organs (usually considered an accessory sex organ).

W

Walrus. Massive aquatic carnivorous mammal having two tusks jutting from upper jaw, thick hide. heavy layer or blubber and very short tail.

Warm Blooded. Homotherm animal whose body tempertaure does not fluctuate with external/environmental temperature- birds and mammals.

Warts. Rough outgrowths on skin as in case of toads. They may be present on palms. fingers etc. in case of human beings

Wasps. Large group of hymenopterous insects (Vespoidea), slender body, thorax and abdomen joined by narrow stalk, majority parasitic on other insects; most common ones are *Vespa* spp.

Water Bird. Swimming game-bird (Aves) : Bird that frequents water.

Water Boa. Anaconda: gigantic South American snake. *Eunectes* murius (Ophidia: Reptilia).

Water Boatman. Aquatic bugs (Cortxidre : Hemiptera) that inhabit lakes and ponds - *Arcocorixa* spp.

Water Buck. Large shaggy gregarious antelope (Kobus spp.) of South Africa: brown with white ring on rump; inhabits swampy plains and frequents river banks.

Water Buffalo. Common domestic buffalo of India *Bos 'buballs* (Artiodactyla: Mammalia).

Water Deer. Small Chinese musk-deer (Mammalia) of aquatic habits.

Water Dog. Any dog of aquatic habits (poodle etc.); valuable to sportsmen in hunting water-fowls.

Water Hen. Moorhen: ralline bird *Gallinula chloropus.*

Water Microbiology. Branch of microbiology that deals with normal and adventitious microflora of natural and artificial water bodies.

Water Mouse. Water - rat or water-vole. aquatic rodent *Arvtcole amphtbtus* (Rodentia: Mammalia).

Water Pockets. Number of pouch-like diverticula from rumen of camel that do not store water as believed. They are actually structures that contain food in the process of digestion.

Water Scorpion. Predaceous aquatic bug (Nepidre: Hemiptera) forelegs raptorial. hind legs adapted for walking; bite painful.

Water Snake. Aquatic and semiaquatlc snakes. having flatttened tail adapted for swimming - Hydrophildre (sea-snakes). Homalopsinre (fresh-water forms) and *Natrtx* spp. (amphibious).

Water Spider. Fresh water spider (Aranelda : Arachnida) that lives temporarily under water supported by film of air trapped around in its body hall'.

Water Strider. Small slender bug (Gerrldre : Hemiptere) with long slender legs. found skating on water surface ; feeds on small insects.

Water Vascular System. Intemal closed system of canals and duets containing fluid. that Is supplied to tube-feet. found only in echinoderms.

Wax. Type of lipid useful in protection of organism. Found in eplcuticle of many insects.

Wax Glands. Tubular glands found In passage of external auditory meatus. They produce an oily secretion that dries up into thin wax-like flakes called wax of ear. This substance prevents dust entry into ear passage. Also called ceruminous glands.

Web. Thin membranous structure joining toes in certain animals for example frog.

Weevil. Beetle (Coleoptera: Hexapoda) with head prolonged into beak or snout, chewing type of mouthparts situated at the end of snout; feeds on leaves and other plant parts including tree trunks and stored products such as flour. grains, tubers etc.

Wenckebach Phenomenon. Repeated sequences of cardiaccycles in the electro cardiogram in which PR-interval lengthens progressively until a ventricular beat is dropped.

Whales. Common name for large sized aquatic mammals like *Balenoptera* indica. They have naked skin. paddle-like limbs and a very thick layer of fat in subdermis called blubber.

Whistling Hunter. Wild dog, *Cuonn alpinas* (Carnivora: Mammalia). instead of barking or howling. this dog only whistles: hunts cheetal. neelgai and wUd pigs for its food.

White Ant. Common name for termite (Isoptera : Hexapoda).

White Matter. Tissue of central nervous system cosisting of myelinated nerve fibres.

Reaction. When pointed object is drawn lightly over the skin the stroke lines become paler. This is called white reaction and it appears in about 15 seconds.

Wigeon. Common name for various wild ducks (Anseriformes: Aves) - *Anas penelope* found in North Eurasia: *A.americana* native tc North America etc.

Wild Type. Normal/standard form 01 a gene or an individual as found typically in nature.

Wwilson's Disease. Due to disturbance of copper metabolism an autosomal recessive physiological disorder is caused.

The body suffers with a severe copper, anaemia and there is no urinary catechlamine excretion.

Wish Bone. Fused clavicles and interclavicle forming a fork-shaped bone articulatign between pectoral girdles and sternum and acts like a spring betweentwo halves of pectoral girdle.

Wolf. Wild gregarious, flesh-eating mammal, cants lupus having brownish-grey hairy body, erect pointed ears and bushy tail.

Womb. Uterus in femal mammals (including woman) wherein young ones are developed and retained till birth.

Wood Chuck. Ground hog. solitary. diurnal. reddish-brown, burrowing marmot, Marmota monax (Rodentia : Mammalia) found in Canada and East U.SA

Wood Cock. Group of migratory birds having long beak. large eyes, mottled plumage, feed at night in springs and muddy ditches; valued as *food-Philohela* minor (North America). *Scolopax rusticola* (European).

Wook Duck. Carolina duck of North America (Anseriformes. Aves) brightly coloured birds that nest in trees.

Wood Lark. Small European lark. *Lullala arborea* having short tail and variegated plumage.

Wood Pigeon. Ring-dove, *Columba palumbus* (Columbidce columbiformes), wild game-bird.

Woolly. Bearing wool: naturally coV'ered with wool-sheep.

Woolly Aphid. *Ertosoma lanigerum* (Eriosomatidce: Aphtdoidea), cosmopolitan in distribution; nymph secretes white woolly filaments of wax over its body; most destructive pest of apple trees all over the world.

Woolly Bear. Large hatty caterpillars o(various mo!hs (Arctudae: Lepidoptera) - *Estigmene lacttnea, Isla isabella, Pericallia ricini* etc.

Wooly Hand Crab. Mitten crab (Crustacea); chinese crab that has caused extensive damage to river banks in Germany during last 77 years.

Woolly Monkey. Small monkey (Platyrrhina : Primates) of Amazon forests (South America). 45 to 50 cm long. ha' l ggblack face, flat nose. grey-brown body with woolly fur and about 60 cm long prehensile tail.

Worms. Common name for small rounded animals like tape-worms. pin-worms, hook-worms, thread-worms etc.

Wren. Small insectivorous singing bird. *Troglodytes troglodytes* and allied species (Passeriformes : Aves) having dull brown plumage and erect. short tail.

X

Xanthine Oxidase. Enzyme that catalyses oxidation of xanthine of uric acid.

X-Chromosome. One of the sex chromosome which is concerned with sex determination. Individuals with double dose (XX) are females while those with single dose and a Y-chromosome (XY) are males (in XX sex determination).

Xenopneusta. Small order of newly erected phylum Echiurida that includes worm-like marine animalcules haVing anterior detachable prostomium.

X-Gland. In some arthropods endocrine gland passin~ secretions along axons to sinus gland at the base of eye stalk.

X-Iinactivation. Consideration and resultant inactivation of all or part of any X-chromosome in cell in excess of one.

Xiphisternum. Flattened cartilagenous structures attached below mesosternum in frog. In rabbit it is attached behind last stenabra.

Xiphoid Processes. Xiphisternal processes: posterolateral boarders of sternum in bi:-ds are expanded to form elongated flattened structures called xiphoid processes.

Xiphosura. Sub-class of chelicerate arthropods to which horse-shore crab belongs.

XY-Chromosomes. Chromosome concerned with determination of sex. In most animals the male most has XY chromosomes and the female XX chromosomes, but in some male, has XX chromosomes; and femal XY-chromosomes. A few cases are known to have XO sex detrmination where male has XO-chromosomes and femal XX-chromosomes.

Y

Y-Chromosome. One of the sex chromosomes in XY sex determiniation system. Individuals with XY are males and those with XX are females.

Yolk. Protein and fat combination in the egg that provides nutrition for developing embryo.

Yolk Plug. During epiboly in from gastrula the entire embryo is encroached upon by pigmented micromeres leaving a small area in blastopore which is callled yolk plug.

Yolk Sac. Extra embryonic membrane which is sac-like in structure and formed mostly in reptiles. birds and egg-laying mammals. It surrounds the entire yolk of the egg and serves to carry out nutrition of the embryo.

Yolk Stalk. Stalk-like stnlcture connecting yolk sac with intestine of embryo in case of *Scoliodon* spp.

Y-Organs. Pair of non-neutral structures in anterior protion of crustacean body and is source of moulting hormone.

Younginiformes. Suborder of Eosuchia, extinct, small l epidosaurian lizard-like reptiles that were found in South Africa during late Paleozoic (Permian) and early Mesozoic (Triassic) periods.

Ypsilophuriidae. Family of Dactylochirotida (Echinodermata); body fusiform or U-shaped; with 8 to 10 tentacles and permanent spire on plates of test.

Yow. Ewe or female sheep.

Z

Zebra. Mrican ass or horse like mammal (Perissodaetyla) with black & white striped body - common species is *Equus burchelli* while *E. zebra* and *E. quagga* are extillct.

Zeis Glands. Sebaceous glands associated with follicles of eye-lashes; their secretion keeps eyelashes smooth and oily.

Zero Nerve. Terminal nerve emerging out from ventral side of cerebrum through the neuropore and passes into olfactory mucous membrane. It is sensory nerve and bears a ganglion mid-way from its origin.

Order Kinetics. Kinetics in which rate of reaction is dependent on concentration of any of its reactants. This would occur if enzyme concentration was the limiting factor.

Zeugloptera. Suborder of Lepidoptera (Hexapoda) formerly placed under Homoneura (= Jugatae); comprises a single family Micropterygidre.

Zeupopodium. Antebrachium forming fore limb and shank in hind skeleton support in former is formed by radius and ulna bones and in the later by tibia and fibula bones.

ZHO. Hybrid domestic cattle found in certain parts of Himalayas; cross between male yak (Bos grunniens) and cow (B *indicus).*

Zoea. Larval stage of certain crustacea - crabs etc.

Zona. Any encircling structure - external or internal.

Zona Glomerulosa. Outer zone of adrenal gland's cortex made of whorls of cells which are continuous with columns of cells that form zona fasciculata. Enzymic mechanisms for aldosterone biosynthesis is limited to zona glomerulosa.

Zona Radiata. Striated primary membrane in mammalian egg separated by follicullar cells.

Zona Reticularis. Innermost zone of adrenal cortex where the cellular columns become interlaced in a net-work.

Zonla Occludens. Tight junctions between epithelial cells, usually having a ring-shaped configuration and serving to occlude trans epithelial extracellar passage.

Zonule of Zinn. Fibres holding crystalline lens of eye in place.

Zoo. Animal: collection of animals.

Zoobiotic. Living in association with another animal. ZOOBLAST: animal cell.

Zoo Chemistry. Chemistry of animal body.

Zoo Culture. Domestication of animals.

Zoodynamics. animal physiology.

Zooerythrin. Red pigment in birds' feathers.

Zoogamy. Sexual reproduction of animals.

Zoogeny. Theory of origination of living beings (animals).

Zopilote. *Urubu*: small American vulture, *Cathartes aura* (Falconiformes: Aves) having reddish head and neck and dark plumage feeds on dead animals.

Zorapetra. Order of little known insects, having 9 segmented moniliform antennre, biting and chewing mouthparts: thorax as long as abdomen and slight or no metamorphosis - only one family (Zorotypidre) with one genus *(Zorotypus)* and around 22 described species.

Zorille. Slender black and white mustel1ne animal, Zonlla species, about 60 cm long with bushy tail another 50 to 60 cm: when disturbed it ejects foul-smelling flUid: found in most parts of Africa

Zorro. Fox-like wild dog (Carnivora: Mammalia) found in South America.

Zygantra. Pair of articular surfaces present at the posterior part of vertebrae in snakes.

Zygapophyses. One of the four projections (two anterior and two poserior) extending from neutral arch of a vertebra. Those of posterior pair articulates with anterior pair of vertebrae next behind.

Zygentoma. Suborder of Thysanura, primitive Wingless insects haVing elongated flattened body: cerci and median caudal filament long and many segmented - silverfish, *Lepisma sacharina.*

Zygodactylus. Foot in birds like parrots etc., wherein first and four digits are directed .backwards third and second are directed forewards to provide with firm grip on the branch of a tree.

Zygomatic Arch. Jugal and process of squamosal form a arch in case of mammals that enclose a single fossa. Ths arch is separates the temporal fossa from orbits.

Zygosis. Sexual union of two unicellurar organism.

Zymogen. Proenzyme: inactive precursor of enzyme; liberates active enzyme on reaction with appropriate kinose. *Cells*: cells in stomach mucosa that secrete pepsin. *Granules*: secretary granules present in salivary glands that contain salivary enzymes.

***